Human Development in its Social Context

Also published in association with the United Nations University:

Transforming the World—Economy?
Development as Social Transformation

Human Development in its Social Context

A Collective Exploration

edited by Carlos A. Mallmann
and Oscar Nudler

contributions by: Telma Barreiro
W. Lambert Gardiner
Bennie A. Khoapa
Catalin Mamali
Eleonora Barbieri Masini
Ian Miles
Ashis Nandy
Gheorghe Paun
Maria Teresa Sirvent
Sulak Sivaraksa

A15042 421364

BF
713
.H95x
West

HODDER AND STOUGHTON
LONDON   SYDNEY   AUCKLAND   TORONTO
in association with the United Nations University

British Library Cataloguing in Publication Data

Human development in its social context: a
  collective exploration.
  1. Developmental psychology
  I. Mallmann, Carlos A.   II. Nudler, Oscar
  III. Barreiro, Telma   IV. United Nations.
  University
  155   BF713
  ISBN 0 340 38517 0

First published 1986

Copyright © 1986 The United Nations University

All rights reserved. No part of this publication
may be reproduced or transmitted in any form or by
any means, electronic or mechanical, including
photocopy, recording, or any information storage and
retrieval system, without permission in writing
from the publisher or under licence from the Copyright
Licensing Agency Limited. Further details of such licences
(for reprographic reproduction) may be obtained from the
Copyright Licensing Agency Limited, of 7 Ridgmount Street,
London WC1E 7AE.

Typeset by Colset Private Limited, Singapore

Printed in Great Britain
for Hodder and Stoughton Educational
a division of Hodder and Stoughton Ltd, Mill Road,
Dunton Green, Sevenoaks, Kent by
Richard Clay (The Chaucer Press) Ltd, Bungay, Suffolk

Contents

Preface ix

Section I   Introduction

Chapter 1   On the Human Development Concept and Contemporary Ideological Systems
Oscar Nudler   3

Section II   The Human Development Concept

Chapter 2   Towards a Model of Human Growth
Telma Barreiro   21

Chapter 3   On Human Development, Life Stages and Needs Systems
Carlos A. Mallmann   50

Chapter 4   On Turning Development Inside-Out or (better) On Not Turning Development Outside-In in the First Place
W. Lambert Gardiner   63

Section III   Human Development: Psychosocial Aspects

Chapter 5   Classification and Hierarchization in Motivational Fields: Co-Evolution Vectors
Catalin Mamali and Gheorghe Paun   93

Chapter 6   Small Groups and Personal Growth: Distorting Mechanisms Versus the Healthy Group
Telma Barreiro   113

Chapter 7   The Development-Adaptation Dialectic
Oscar Nudler   126

Section IV Human Development: Regional Perspectives

Chapter 8	Human Development in Contemporary Industrial Societies Ian Miles	141
Chapter 9	Human Development and Childhood Eleonora Barbieri Masini	179
Chapter 10	Human Development and Popular Culture in Latin America – Case Studies Maria Teresa Sirvent	190
Chapter 11	The African Personality Bennie A. Khoapa	213
Chapter 12	Buddhism and Development Sulak Sivaraksa	233
Chapter 13	The Idea of Development. The Experience of Modern Psychology as a Cautionary Tale and as an Allegory Ashis Nandy	248
Index		261

About the Contributors

Telma Barreiro, Lecturer and Researcher, working actively in designing and applying group techniques in courses for teachers.

W. Lambert Gardiner, Research Associate, GAMMA, Université de Montréal-McGill University.

Bennie Khoapa, Professor at Case Western University, USA.

Carlos A. Mallman, former Executive President of Fundación Bariloche, Argentina; Coordinator of the Goals, Processes and Indicators of Development project of the United Nations University, Tokyo; Co-Director of the Center for the Study of Human Development associated with Fundación Bariloche.

Catalin Mamali, member of the Research Center for Youth, Bucharest.

Eleonora Masini, President of the World Futures Studies Federation; Coordinator of the project 'Household, Gender and Age' for the United Nations University, Tokyo; Lecturer of an inderdisciplinary course in Social Forecasting, Pontificial Gregorian University, Rome.

Ian Miles, member of the Science Policy Research Unit, University of Sussex.

Ashis Nandy, Senior Fellow of the Centre for the Study of Developing Societies, Delhi; Chairman of the Committee for Cultural Choices and Global Futures, Delhi.

Oscar Nudler, Co-Director of the Centre for the Study of Human and Social Development associated with Fundación Bariloche; Coordinator of the Economic Aspects of Human Development project of the United Nations University, Tokyo.

Gheorghe Paun, main Researcher in the Division of Systems Studies, University of Bucharest.

Maria Teresa Sirvent, Director of the Institute of Education of Buenos Aires University; Researcher in the area of leisure time and non-formal education (especially in marginal groups).

Sulak Sivaraksa, Program Chairman of the Siam Society; Coordinator of the Asian Cultural Forum on Development.

Preface

This book is an output of the Human Development Study Group of the UNU-GPID Project. It is based mainly on the papers presented at the meeting of the Study Group, held at Fundación Bariloche, San Carlos de Bariloche, Rio Negro, Argentina, 11-15 December, 1980, and the discussions which took place during the meeting. The final versions of the papers took these enlightening discussions into account. In this sense, this book is a Collective Exploration.

Section I, written by O. Nudler, shows the multifaceted nature of the key concept of this volume – human development – and reviews in its light some of the major contemporary ideological systems. Thus, we consider this chapter as a general introduction to the problem which concerns us here, that is, human development in current social conditions.

Section II – a theoretical examination of different aspects of the human development concept – consists of three chapters. In the first one T. Barreiro, after arguing for the notion of *continuous* personal growth ('primary growth' is to be continued by 'maturity growth'), makes a detailed comparison between 'the successful person, well-adapted to utilitarian culture' and 'the person who has achieved maturity and strength through his maturity growth'. In the second chapter of Section II, Carlos A. Mallmann brings into focus the link between Erikson's theory of life stages on the one hand, and potentialities and needs on the other, linking in this way two fields usually disconnected. He maintains that good development 'consists of solving the sequence of nuclear crises positively'.

The previous two chapters emphasize, in contrast to positivist approaches, the non-neutral character of the concept of human development. It is assumed, rather, that full human development is at the same time good development, positively valued on ethical standards. In the last chapter of this Section, W. Lambert Gardiner looks at the mal-development/good development distinction from yet a different angle, namely, the dispute between behaviourism and humanism in psychology. He sees these two approaches as implying opposite views on human development: the 'outside-in' and the 'inside-out', respectively. He describes the main assumptions behind both views in considerable detail and argues extensively for the latter.

Section III, which focuses on psychosocial and personality aspects of human development, starts with C. Mamali and G. Paun's chapter on motivational fields, of which they outline a mathematical model. An important dimension in connection with motivational fields stressed by the authors is personal development considered as *codevelopment*, that is, development as a function of personal interaction. This dimension is also taken up by T. Barreiro in the next chapter, although from a different standpoint. She explores the mechanisms which distort human communication and interaction and, in contradistinction to them, she describes the 'healthy group'. A review of the effects on human development of both social environments is proposed. The third Section ends with a paper by O. Nudler which deals with the different ways in which people react to the social pressure to adapt. In this connection the author distinguishes among three main personality types characterized respectively, by over-adaptation, failure in adaptation and transcendence over adaptation. He describes the bearings of each of these types on human development.

Section IV is devoted to a variety of regional perspectives and to analysing and assessing human development in different regions and diverse economic and social conditions. The first chapter, by I. Miles, makes, on the basis of recent statistical data, a detailed and critical survey of the way in which the main activities and associated institutions of industrial societies, such as remunerated work, schooling, leisure and so on, impinge upon human development or 'agency'. The next chapter, by E. Masini, also addresses the problem of human development in industrial society but from a quite different angle, namely, the visions of the future expressed by children living in contrasting Italian neighbourhoods.

The third chapter of Section IV, by M.T. Sirvent, is the result of field studies undertaken by the author in two communities located in the outskirts of Buenos Aires and São Paulo, respectively. In this chapter, the conditions for human development in poor, marginal Latin-American settings are seen as an interplay between prevailing dependency and seeds of autonomy, between authoritarianism and participation. The next chapter, by B.A. Khoapa, shows the negative impact that the imposition of an exogenous model of human development – the Western model – has had on the African Personality. The author argues, therefore, for the need for reaffirming the African Personality with its associated cultural values as a necessary pre-condition for full human development in black Africa. The final two chapters are written from an Asian perspective. S. Sivaraksa, from Thailand, opposes the Buddhist conception of human development to the Western conception and reviews the undesirable effects of this latter on human development. The closing chapter, by A. Nandy from India, starts reflecting on the development/underdevelopment distinction and on what he contends is its usually wrong interpretation in terms of health/ill-health or normality/abnormality. After supporting the idea of taking the distinction as a continuity and not as a dichotomy and after arguing for the convenience of looking at it from the health extreme instead of taking the ill-health extreme, as is usually done, Nandy applies his analyses outside psychology and proposes to look at a nation's underdevelopment 'as a normal or healthy response to oppression, humiliation and inequity of cultures'.

As can be concluded from the foregoing, rather sketchy description, this book contains a rich variety of studies around its organizing focus, that is, human development in its social context. Perhaps the main shared feature throughout the different chapters is the use of holistic, non-individualist and value-laden concepts of human development as critical tools to describe and assess the human condition in the world in which we live and to outline alternatives to it.

Before ending this preface we would like to thank whole-heartedly all the contributors for their unfailing co-operation and understanding without which this book would not have been possible.

C.A. Mallmann and O. Nudler

Section I
Introduction

1
On the Human Development Concept and Contemporary Ideological Systems

Oscar Nudler

The main purpose of this introduction is to show the potentiality of the key concept of this volume – human development – as a tool for social analysis, critique and planning.

In the first section of this chapter some general reflections on the human development concept are made, just to give an idea of its complexity. The focus is then laid on a pair of its dichotomous dimensions, namely, internal–external on the one hand, and individual–collective on the other.

In the second section, these dichotomies are applied to the analysis and assessment, from the human development perspective introduced here, of the dominant ideological systems of our times, that is, liberal capitalism and state socialism. The analytical effort is concentrated here on ideological systems as they actually work in the real world rather than on the doctrines of their founding fathers or followers. As is well known, the gap between both of them is usually rather wide.

In the third section the focus shifts towards alternative humanistic models of human development, particularly those grounded on humanistic psychology and states of consciousness research.

Finally, in the last section, after summarizing briefly the main conclusions out of the preceding analysis, a tentative list of requirements that a human centered society should meet is proposed.

1 The human development concept

To approach the human development concept in its connection with the social context, it might perhaps be useful to follow an indirect path and consider first a very common application of the term 'inhuman'. It is rather usual to hear people who do not feel quite happy with a given social order accusing it of being 'inhuman'. Widely different regimes, sometimes opposite to each other, have been the target of such an accusation. The reasons, of course, are different in each case: blatant human rights violations, gross social inequity, frantic consumerism, spiritual vacuum and so on. In spite of these disparities, the different uses of our term still seem to have a common meaning core. In all cases in which a social and political order is considered inhuman, what is probably conveyed is that such a regime does not allow people living within it to maintain or develop their human-ness, where 'human-ness' refers to what is more proper to human beings, to what differentiates them from other entities populating our universe such as machines, animals, fossils and so forth.

The implied puzzling expression 'the development of human-ness in human beings' does not seem to point to any particular feature like physical strength, intelligence or moral sense, but rather to the whole set of powers which characterizes a human being. In other words, 'the development of human-ness in human beings' or, in short, 'human development', is used, at least implicitly, as an *holistic* concept. Sometimes, to make this characteristic explicit, the term *full* human development is introduced.

We could say thus that when a given social order is called 'inhuman', a contradiction between it and full human development is implied.

But what is the meaning of 'full human development'? The holistic nature of this concept makes any attempt to reach a satisfactory definition of it particularly difficult. It could be said, for example, that full human development is a process which leads to holistic health or maturity, but then the weight of the definition is just transferred to 'health' or 'maturity', concepts which are no less difficult to define or explicate.

To take a more elaborated example of such difficulty let us look at the definition of healthy personality proposed by D. Schultz[1] on the basis of the common points he found by comparing seven personality theories (those of C. G. Jung, E. Fromm, A. Maslow and others):

1. Healthy persons are capable of consciously, if not always rationally, directing their behaviour and being in charge of their own destinies;
2. such persons are aware of their strengths and weaknesses;
3. they have a view towards the future but without substituting the future for the present;
4. psychologically healthy persons do not long for quiet and stability but for challenge and excitement, for new goals and new experiences.

Taken as a general definition or clarification of health, full develop-

Keep the office cleaned (not sure stuff)
hang up cottons & crayons
mark food in fridge
use up left-overs
Groups to be done w/ both staff
Make dinner during FTS
FTS — needs to be signed by both staff & OT — when homework is done

NOTES

Ethnic Succession:

One gp replacing another in neighborhoods, jobs + schools.

Poorest → Poor → lower/middle → middle →
Unskilled Skilled Blue Collar White Collar

Harlem: Jewish / S. Blacks / Barbados - Jamaica / Puerto Rican / Ethiopian + El Salvador ment or maturity, Schultz's list could raise various objections. For instance, it seems to imply a purely individualistic view (social interaction or human development understood as *codevelopment* are not mentioned). Furthermore, the list looks too Western (especially as revealed by item 4). However, Schultz's definition contains no doubt valuable insights and it is debatable, anyway, whether a universal definition of full human development, maturity or health is possible or desirable. What is in any case clear is that the wide scope and complexity of these concepts make possible different readings of their rich, open meaning.

I do not propose then to give a new definition of human development. Rather, what I intend in what follows is to focus on the two pairs of dichotomous dimensions mentioned above and, particularly, on the 'isms' which derive from a one-sided emphasis on one of their extreme poles, that is individualism–collectivism on the one hand, and externalism–internalism on the other. Individualism is the view according to which each human being is a separate entity, not only in the obvious physical sense, but also mentally and spiritually. An immediate consequence which follows from this basic assumption is that any individual has a private world which is unique and, in the last resort, uncommunicable. Conversely, collectivism rests on the belief in an essential likeness among human beings or, more precisely, among the members of a given class, nationality, race, religion or any other preferred division. The notion of private worlds which are qualitatively diverse is usually seen from this perspective not only as false but also as dangerous.

Our second polarity – externalism–internalism – can be characterized on the basis of what counts as a criterion for assessing human development. According to externalism, the development of a human being is directly correlated to the amount and importance of his achievements in the external world. On the contrary, what matters for internalism are internal states, feelings, capacities, not necessarily manifested through achievements in the external world.

In the next section, these brief characterizations will be completed through the application of our dichotomies to the description of the way in which the major dominant ideological systems of our times envisage human development. I should, however, anticipate that the criterion of full, good or adequate human development preferred here has mainly to do with a dynamic balance between the opposite poles of the pairs referred to above, i.e. achievements in the external world and inner growth on the one hand, and high social integration and preservation of individual originality on the other. This criterion could be seen as universal but only in a formal way since the dynamic balance referred to can probably take an infinite number of diverse cultural and personal forms. As is shown in the next section, present-day dominant ideological systems can be located in the extremes of each pair. That is why they are considered, from the perspective adopted here, as more opposite than favourable to full human development.

2 Dominant models of human development

The ideological systems to be addressed in this section are historically connected with the impact of what social scientists usually call 'modernization processes'. The so-called 'modern society', which is the result of such processes, can be characterized in different ways, according to the relevant aspects which are picked up. In view of the purpose of this study, I shall focus on the nature of human relations in modern society. In this connection, the traditional distinction in the social sciences between primary relations (face-to-face, involving the person as a whole) and secondary relations (non-personal, within the limits of specific roles) is still helpful. However, I shall depart slightly from this categorization and use 'primary' in the narrow sense of 'intimate', 'very close' (as is supposed to happen in the nuclear family case) and introduce a third category – *community relations* – between the primary and secondary ones. Community relations are not supposed to be as intimate as nuclear family ties but somehow still keeping the personal and holistic flavour of primary links. These relations have their natural locus in intermediate social groups, that is, groups whose size and organization structure make possible personal contact among all their members.

Now vis-à-vis this threefold classification, modern society can be characterized as having a built-in tendency to increase the social space covered by secondary relations at the expense of the other two types of relations. As to primary relations, not only the size and number of functions of the family diminish with modernization, but also internal family links become increasingly unstable. Community relations, in their turn, are, as already shown by Tönnies, Weber and others a long time ago, the favourite victims of modern society since they tend to disappear completely.

This categorization of human relations in modern society, together with the two pairs of dichotomous categories introduced in the preceding section, make up the conceptual framework to be used in what follows. The ideological systems to be considered here are, as mentioned at the beginning, liberal capitalism and state socialism. They are grouped together, first, because the two of them are in a dominant position in a large part of the world and, second, because they share a basic confidence in the modernization paradigm, particularly industrialism.

The conceptual scheme to be followed in the next two sections can be represented by means of a 2 × 2 table:

Table 1.1 *Current models of human development*

	individual	collective
external	Liberal Capitalism	State Socialism
internal	Liberal Humanism	Communalism

From the perspective argued for here, a desirable model of human development would not be located in any of the cells but rather in the centre of the preceding matrix.

The liberal capitalist model

To start with the individualism-collectivism distinction, it is hardly debatable that the classic liberal capitalist model is inclined towards the individualist pole. Leaving aside the old issue about which came first, the capitalist system or the individualist attitude, it is an obvious historical truth that liberal capitalism and individualism have been closely associated with each other. As a matter of fact, the belief in the radical separateness of individual consciousness has been a central tenet of Western philosophy throughout all its history. Averroes' interpretation of Aristotle, one of the few examples in the opposite direction, had only a fleeting influence on Western thinking. In fact radical doubt about the existence of the physical world and other minds – a characteristic theme of modern philosophy – was grounded, as is clear in Descartes and other modern philosophers, on the certainty about the separate existence of one's own mind. Such a view of the human being as a basically isolated entity extends far beyond the boundaries of philosophy to underlie various influential metaphors used in Western psychology and social science, such as Freud's psychic apparatus or Parson's social actor. It also stands, of course, behind the *homo oeconomicus* fiction of classic economics and plays an essential role in liberal political thinking. The questions posed by John Stuart Mill at the beginning of his chapter on 'the limits of the authority of the society over the individual' are quite revealing:

> What then is the rightful limit to the sovereignty of the individual over himself? Where does the authority of society begin? How much of human life should be assigned to individuality and how much to society.[2]

Here we have the individual on the one side and society on the other. The classic problem before liberal political thinkers consists of devising a compromise fair enough for both parts.

The consequences of individualism for human development are

various and have been abundantly pointed out. On the one hand, the emphasis placed by liberalism, particularly in its expansion phase, on individual rights, cannot be overlooked. It is difficult to deny the historical contribution to the idea of human development made by this liberal view but, on the other hand, other characteristics of the liberal credo, linked to its individualist assumption as well as to externalism, have had a much more dubious impact on human development. It suffices to mention in this connection the stress put on individual success and the associated thrust towards competition in all spheres of life such as work, sports, sex and so on. A classic paradox used to legitimize the high value attached to competition is that general well-being can be attained only if people work with enthusiasm in favour of their private interest. The point of mentioning this liberal paradox here is not to discuss its truth value or applicability, dubious enough in a world situation characterized by economic crisis and high concentration of power. What I want to stress is its inertia as an ideology, particularly in the middle classes, a fact which is probably due to its close relationship with the character structure associated to liberal capitalism. Such ideological strength of the competition idea explains why its decreasing applicability goes mostly unnoticed for middle classes. A related assumption is that although the particular way in which each individual chooses to develop belongs to their private domain, the liberal capitalist state must and can, in contradistinction to other systems, assure everyone the right to try freely the selected alternative. However, in fact only a small élite has a wide range of opportunities for exercising free will in the social arena. The cases of those who have been able to attain success starting from low social levels are wielded as an evidence that the reward really exists, that social mobility is possible. As to the rest of people who in spite of that do not manage to succeed or just maintain their initial position, the classic liberal answer is that they should blame themselves for their failure. But the problem is that due to unemployment, recession and related phenomena showing the end of an economic expansion which was once assumed to be unlimited, this 'rest' is a growing portion of the population of industrial societies. Ironically, instead of revising the assumptions of the liberal model vis-à-vis this reality, mainstream economists and politicians, and not only those belonging to the so-called 'new right', attribute the difficulties to deviations or distortions in the application of the model and not to the model itself. As a result, an increasing tendency to leave aside socialist or Keynesian policies and a shift to more orthodox liberalism is presently in vogue. Even though such a shift has not produced, generally speaking, the expected results, this failure does not seem to threaten seriously, at least in the short run, the popularity of liberal leaders and ideas. This could be seen as a nice illustration of the inertia enjoyed by ideologies which have been once 'progressive' and have managed to shape society according to them.

Turning now to our second analytical dimension, it is evident that liberal capitalism is inclined towards externalism. The main component of externalism is, as mentioned above, the idea of achievement in the external world and in connection with it the high appreciation of success. Typical examples of highly developed people might be, according to the model and, within it, according to varying personal and cultural preferences, businessmen or managers who have proved their capacity to expand sales and profits in a difficult market, politicians who are able to seize power and keep it thanks to their knowledge, courage and/or intuitive abilities, scientists who have contributed significantly to the advance of knowledge, particularly if they are awarded the Nobel prize, famous writers, singers, movie stars, football or tennis players and so on. It is admitted, however, that success does not always automatically accompany achievement. In some unfortunate cases success may never come in the life span of a person who deserves it. In this case only posterity could repair the injustice. At any rate, success, in the short or in the long run, during or after life, is strongly associated by the liberal model to achievement and full human development.

As argued by Max Weber[3] in his classic work on the relationship between protestantism and capitalism, protestant ethics played an essential role in the origin of capitalism, particularly in view of its emphasis on external achievements taken as signs of salvation. And if protestantism was the main religious source of the externalist thrust built in early capitalism, behaviourism has been, in our century, one of its more significant secular sources. W. L. Gardiner,[4] in his presentation of the main assumptions about human nature which underlie behaviourism, has made clear the connection between such assumptions and externalism:

1 The person has only extrinsic needs.
2 The person is conditioned from outside in.
3 The person has only extrinsic worth.
4 The person is an interchangeable part.
5 The person has contractual relationships.

As we can see, behaviourism has taken externalism to an extreme by denying even the existence of inner life and, therefore, the distinctive value associated to it by traditional spiritual philosophies and different branches of humanism. The substitution of orthodox behaviourism by more sophisticated approaches like cognitive psychology or artificial intelligence research, both of them based on the computer root analogy, has left untouched, however, the behaviourist view of man as a machine.

Externalism, at least in its extreme behaviourist form, collides with individualism as long as the latter affirms the irreducible originality and non-interchangeability of human beings. There is a tension built in the liberal capitalist model, an unresolved tension between freedom and determinism, inner autonomy and external manipulation. Such tension could now be solved but in favour of the second terms of the just mentioned pairs. Nobody can assure, of course, that nightmares in the style of *Brave New World* will come true one day. But what can be asserted with relative certainty is that the assumptions of the classic liberal capitalist model have lost any positive, creative impact on human development they could have had in the past. However, the shift from the extreme individualist pole to a more social one as performed by democratic socialism in Western Europe, is facing growing difficulties, particularly, though not only, on the economic front. As a consequence, orthodox liberalism is emerging as the dominant force once again.

The above implies, from a human development point of view, that a substantial revision and not a mere adaptation of the assumptions of the liberal capitalist model is needed. As a matter of fact, such revision is already under way. The discussion in section 3 of the humanist model of human development is to a large extent inspired by it.

The state-socialist model

According to our above-mentioned purpose of focusing on working ideologies, Marx's original human development model will not be discussed here but rather the model which is actually followed in the socialist countries. The official doctrine takes for granted that Marx's model has been implemented in the USSR but for many Marxist thinkers the truth is just the opposite. A more balanced approach would perhaps recognize that socialist countries have applied Marx's ideas in some aspects but that they have also departed considerably from them in others. This ambiguity could be somehow traced back to the relatively ambiguous nature of Marxism itself, as is suggested by the contraposition between a young Marx and an old Marx, between the *Manuscripts* and *Das Kapital*. E. Kamenka[5] summarized such antimony as follows:

> The intellectual crisis which is currently observed in the social democratic movement is . . . a crisis which originates from the tension between Marx's emphasis on economic rationalism and material supply on the one side, and his emphasis on an authentically human moral which overcomes the very notion of ownership and the divorce between means and ends on the other.

Whatever the relation between classical Marxism and state socialism might be, it is clear that the latter is, like liberal capitalism, definitely externalist. The inclination, even obsession, towards material achievements, especially if they imply the application of hard technology, is also at the core of the model. The ideology of industrialism is the dominant ideology and it goes hand in hand with an externalist concept of human development. The great difference between state socialist and liberal capitalist models of human development lies, needless to say, in their attitudes towards individualism. While liberalism explicitly adopts and encourages individualism, state socialism explicitly rejects it. The main proclaimed goals – the strengthening and growth of socialist states and the achievement of a communist society – are contrasted to individualism and subjectivism, 'bourgeois residues' to be eliminated. In this model the degree of personal development is a direct function of the degree of identification with and contribution to social development as officially defined. In contrast to liberalism, personal development is not seen as a mostly private affair. That is why the socialist state intervenes much more freely and openly than the capitalist state in personal areas such as the determination of the place of work, residence, studies to be undertaken and so on.

Now socialist countries have not been an exception to the weakening and even complete disappearance of community links, phenomena which are typically related, as mentioned above, to modernization processes. Kolkhozes are not really counter examples to this trend since they are not self-organized and governed but, as the rest of social organizations, they are managed or at least controlled from the top. Generally speaking, the growing dominance of secondary (rational, bureaucratic) organizations is similar to what happens in liberal capitalist societies. The main difference lies in the extraordinary concentration of power in just one of these secondary organizations: the Communist Party. To sum up the preceding description of the state socialist model, we could say that this model is essentially a blend of externalism and bureaucratic collectivism. As regards collectivism, it should be added that due to the narrow space left for participation and creativity not strictly in line with the values of the system, some of the most criticized features of 'bourgeois' individualism, like competitiveness, crude materialism and, in general, lack of social solidarity reappear very intensely, in sharp contradiction to official rhetoric about the end of capitalist attitudes and practices.

In connection with externalism an additional point worth mentioning is that state socialism sticks, like liberal capitalism, to a mechanical, outside-in view of the human being. It is certainly not by chance that Pavlovian reflexology has been highly influential in the rise of American behaviourism.

As argued in the preceding section, liberal capitalism no longer provides an acceptable alternative for human development in current world conditions. The same conclusion holds for state socialism. As in the case of liberal capitalism, its externalist thrust, in the frame of an ever-expanding material growth logic and increasing exploitation of nature and society, is reaching ecological, social, political and human limits beyond which the system turns out to be quite incompatible with human development. And in addition to externalism, the bureaucratic and authoritarian nature of the state socialist model puts further severe restrictions to the full expression and realization of human potentialities.

The general conclusion to be drawn here is therefore that none of the dominant models of human development is satisfactory, at least when they are looked on from the vantage point of a humanist model which stresses, as mentioned before, the need for a dynamic resolution of conflicts between opposite development poles. But what can be said about humanist models of human development? Let us focus on them.

3 Humanist models of human development

Before turning, however, to humanist models some features shared by the two main variants of the so-called *modern paradigm* should be underlined. This will complete a reference frame against which current humanist alternatives can be contrasted.

As has been mentioned, a characteristic feature of the modern paradigm is its one-sided emphasis on external achievements, particularly in the economic and technological domains. Industrial revolution(s) and the decisive impact of science, technology and the type of rationality associated with them in everyday life, have brought undeniable advancements but, at the same time, have contributed to create a world characterized by extreme tensions in almost every area of human life. If we look at the society-nature relationship, the destructive action of industrialism over the environment has been so stressed that it hardly requires any further comment here.[6] Concerning international relations, it suffices to mention the well-known fact that nuclear 'progress' has given the super-powers the potential for destroying each other and the rest of the world many times. As to social relations, let us only recall that although, on the one hand, modernization processes have led to the end of feudal, authoritarian structures, particularly through the building up of national states and their progressive transformation into more open societies, on the other hand the weakening or sheer destruction of community links and, in general, all forms of direct participation of people in social decisions, have created a gap between the level of ever growing macro institutions – transnational corporations, the state, the party – and the micro and individual levels. This lack of *organic* channels of social participation lies in the origin of modern forms of authoritarianism exercised by the new élites and diverse manifestations of contestatory violence as well, two sides of the same coin.

In connection with the impact of modern society on the intrapsychic structure, many studies by Western and Eastern critics of modernity could be quoted here, ranging from Erich Fromm to Gandhi. Let us only mention, however, the fact that social statistics reveal, if a sufficiently long span is taken, a significant correlation between modernization and the increase in psychic and psychosomatic disorders, criminality, drug abuse, suicide rates and so on.

Current critics of the modern paradigm can be divided in different groups, sometimes opposed to each other. Thus, for instance, M. Friberg and B. Hettne[7] distinguish between traditionalists, marginated and 'post-materialists'. Traditionalists are those who resist modernization in the name of a non-modern way of life that they want to defend or restore. The sources of this resistance may lie in non-Western religions and civilizations, old nations and tribes, local communities, peasants, women, informal economies and so on. Marginated people are those who have been unable to find a place in or have been expelled from the modern sector. The unemployed, youth, minorities, non-skilled workers and women are the usual candidates for this category. Finally, 'post-materialists' are those who reject externalist models of human development in favour of a model which emphasizes inner growth and symbiotic relation with the natural and social environment.

Focusing now on humanism, we should distinguish between several forms of humanism, most of which take their name from the doctrines by which they are inspired: Christian humanism, Buddhist humanism, Marxist humanism and so on. Some are, strictly speaking, combinations of humanism and traditionalism, as in the case of the religious variants of humanism just mentioned or, to take another example, Gandhism. In other cases we have a blend with the modern paradigm, as happens with Marxism or with 'scientific' humanism. In what follows only a secular, though not narrowly scientific variety of humanism will be considered. This form of humanism has been developed in the West since the end of the Second World War and is grounded on what is currently known as 'humanist psychology' (inspired both in Western sources, such as European phenomenology and existentialism on the one hand, and consciousness research on the other, and in Eastern sources, especially Buddhism and Sufism).[8] This form of psychological humanism – just *humanism* from now on – is characterized by an explicit rejection of externalism and a pronounced shift or turn towards the 'inner self'. (In some cases even the inner self is surpassed in favor of 'transpersonal conscience'.) What does this 'shift' or 'turn' mean for humanists? The answer to this question will be divided into three topics although it should be borne in mind that the three of them are intimately connected: (1) The humanist view of experience; (2) The humanist view of understanding; (3) The humanist view of human development.

To start with the first topic, let us recall that the main humanist critique of externalist models is that they privilege within the realm of human experience those sectors which have to do, on the one side, with analytical intelligence and, on the other, with mechanical habits. Among the forgotten domains are those linked with the potentialities or abilities to:

(a) experience the world in non-stereotyped, intense ways;
(b) be in contact, recognize and deal non-defensively with deep feelings and drives.

Concerning item (a) humanists claim that the ability to perceive the world without *rubricizing* it, that is, without applying to it a rigid straitjacket, has to be recuperated in view of the atrophy of such ability which usually takes place in externalist cultures due to the prevailing instrumental attitude towards the external world and one's own self. Such reappraisal of direct experience is at any rate a necessary condition for reaching those sources of growth which A. Maslow has called 'peak experiences'.[9]

As to point (b), humanists think that externalism has had the effect of destroying the unity of the human being by dissociating him into body and mind, present and future, reason and emotion. Freud put his finger on this last dichotomy but according to humanists he made the mistake of considering it as an inherent trend of civilized man instead of confining it to the human condition in externalist cultures. Contrary to Freud, civilization and inner unity can for humanists be made compatible.

The second point mentioned above – the humanist view of human understanding – can also be described as a plea for recuperating a denied or largely underestimated ability, namely, the ability to develop a holistic understanding of reality.[10] Gestalt psychologists have already claimed that human beings in their perception of objects and events around them do not normally behave as Cartesian thinkers or digital computers which analytically process bits of information, but apply rather a different ability, that is, the ability to grasp meaningful wholes without going through a painstaking, atomistic analysis in elementary parts. According to humanists, this is not a shortcoming, an approach only acceptable at a pre-scientific level. On the contrary, they argue for the extension of holistic capacities to new domains, including the scientific one. This does not imply leaving aside analytical understanding. It only implies that both analytical and holistic abilities should be integrated so that an 'extended awareness' can be attained.

The humanist theory of human development flows almost entirely from the humanist views of experience and understanding just referred to. However, it is interesting to mention explicitly in this connection the so-called self-actualization theory. According to A. Maslow and other humanist psychologists, the need for self-actualization is a fundamental human need that can be expressed and satisfied only after other needs, more basic in the needs hierarchy, are satisfied. In addition to its unique position in the needs hierarchy, self-actualization has further differences with other needs. Thus, for example, it cannot be explained by means of homeostatic or drive-reduction models such as those used both by Freudian psychoanalysis and behaviourism. These models are probably adequate for biological needs, characterized by a periodical return to a fixed equilibrium point, but self-actualization implies a process which reaches ever new qualitatively different levels. Human development is thus conceived by humanists as occurring in stages which are grounded on previous stages but that, at the same time, are genuinely emergent with respect to them.

Looking now at the humanist model of human development just sketched from the perspective adopted here, its strong internalist tendency on the one hand, and its individualist thrust on the other, are clear. And since individualism is a central feature of liberalism, it does not seem inadequate to call this model 'liberal humanism'. However, it would be somehow unfair to leave the matter at this point. In the first place, humanist psychology is closely associated, as is well known, with the encounter groups movement. It is difficult to deny that the sense of community which encounter groups try to foster is basically at odds with the individualist ethics of achievement and success (although the influence of cooptation processes cannot be ignored).[11] Secondly, there are alternative communities in the search for a humanist way of living. But as far as encounter groups are concerned, they remain outside or parallel to everyday social life, to actual organization of work, education, leisure and so on. And existing alternative communities are still, no matter how intrinsically valuable they might be, at an experimental, very small scale level, devoid of any real weight in present societies.

It can be concluded, therefore, that Western humanism has not managed so far to devise an alternative human development model capable of working in a real, complex society, even though some elements have been set out. What is the explanation for this lack of an articulated and operational humanist alternative? The simplest answer is maybe that many humanists just do not question the liberal social system as a whole. They are interested only in the micro-psychological level and do not relate this level to the macro-social one. To call them liberal humanists would not be so wrong after all.

It is also true, however, that there are humanists who really care about developing a comprehensive humanist social alternative. They usually rely for this purpose on the process of gradual dissemination of small groups and communities based on humanist ethics. But it seems unlikely that such a process can bring about social change if it is not accompanied by a simultaneous transformation of power structures which penetrate people's everyday life-style. If production, education, mass media etc. are dominated by an externalist value system, encounter groups can produce, at most, split personalities living in opposite worlds. And surely this is not a healthy model of human development.

To conclude, the form of humanism referred to above contains valuable elements for a model of human development in the contemporary social context but it is still not a truly workable alternative. Perhaps by integrating creatively liberal humanism, particularly its shift from the external to the internal with elements deriving from other forms of humanism more concerned with actual social forces like, for instance, some manifestations of socialist humanism or the so-called *green movement*, such a truly workable humanist alternative can eventually emerge.

Summary and coda

I have dealt so far with three main models of human development, namely the liberal capitalist, the state socialist and the humanist models. I have used as a reference frame for this review two main dichotomies: external-internal and individual-collective. We have seen that the liberal capitalist model is externalist and individualist and that the state socialist model is externalist and collectivist. Therefore, both systems share externalism, that is, an overemphasis on external achievements and on the means conducive to them, particularly hard science and technology and expanding, vertical power structures.

From the perspective held here this is considered as a stumbling block for full human development. We have also seen that humanist growth theory is internalist and individualist but with some steps taken in the direction of communalism. It was claimed that for a number of reasons this form of humanism had not so far produced a workable social alternative. Since something similar can be said of other varieties of contemporary humanism we might conclude that we are facing a crisis situation with respect to human development models. Maybe this is the clearest sign that we are in the middle of a major historical transition, comparable perhaps to the fifth century AD or the fall of the Roman Empire.

According to what has been already mentioned, an acceptable human development model in the context of modern society should leave room for the simultaneous unfolding and mutual reinforcement of different facets of the human being, that is, inner growth and external achievements, individual uniqueness and social integration, primary, community and secondary social relations. The question which arises now is what are the requirements that a society which promotes such a form of human development should meet. This is not a question which can be dealt with in detail here but a preliminary sketch of an answer can, however, be attempted. A society which favours human development as just defined or, in short, a *human-centred society*, should satisfy at least the following requirements:

1 *Social equity*. Human development is equally possible for all the members of the society.
2 *Inter-regional and international equity*. The society permits and promotes the human development of its members with respect for the integrity of other societies (i.e. no economic exploitation, political domination and/or cultural oppression which prevents the members of those other societies from achieving *their* human development).
3 *Living presence of the future*. Human development of present generations is not to be pursued at the cost of endangering the human development of future generations. I refer especially to the preservation of the natural environment but also to the respect for (not submission to) historical achievements and values which help to define people's cultural identity.
4 *Sensitivity to the present*. Nevertheless the development of the future generations must not be at the cost of the imposed deprivation of the present generation. Construction of a human future is (as implied in point 3) a condition for a human-centred social development process, but oppression of people in the name of a distant future could not be justified from a human development point of view either.
5 *Participation and meaning*. Beyond the preceding principles concerned mainly with different aspects of equity, a human-centred society provides a meaningful frame for human existence so that its members share common feelings and goals and have the opportunity of contributing to their realization without losing personal freedom.

All this could sound hopelessly Utopian. But even if this were the case, a comforting thought to which we could resort is that the ideologies which shape our present world have been once just Utopias, that is, only lights which guided our ancestors through the darkness of history. We badly need a similar guide today.

Notes

1 Schultz, D., *Growth Psychology Models of the Healthy Personality*, New York: Van Nostrand, 1977.
2 Stuart Mill, J., *On Liberty*, New York: Library of Liberal Arts, 1956.
3 Weber, M., *The Protestant Ethic and the Spirit of Capitalism*. Translated by Talcott Parsons. London: Allen & Unwin, 1930; New York: Scribner, 1958.
4 Gardiner, W.L., *'On Turning Development Inside-Out'*, this volume.

5 Kamenka, E., 'Marx's Humanism and the Crises of Socialist Ethics' in Erich Fromm (ed.) *Socialist Humanism*, New York: Doubleday & Company Inc., 1960.
6 See, for instance, Jackson Davis, W., *The Seventh Year, Industrial Civilization in Transition*, New York: W.W. Norton, 1979.
7 Friberg, M. and Hettne, B., 'The Greening of the World. Towards a Non-Deterministic Model of Global Processes', Gothenburg: UNU research papers, 1982.
8 See, for instance, Ornstein, R., (ed.), *The Nature of Human Consciousness*, San Francisco: Freeman, 1973; Tart, Ch., *Altered States of Consciousness*, New York: John Wiley & Sons, 1972.
9 Maslow, A., *Motivation and Personality*, New York: Harper & Row, 1954 (revised edition, 1970); *The Farther Reaches of Human Nature*, New York: The Viking Press, 1971.
10 See, for instance, Stevens, A., *Archetype. A Natural History of the Self*, London: Routledge & Kegan Paul Ltd, 1982.
11 See Blanchard, W.H., 'Encounter Group and Society' in L.N. Solomon and B. Berzon (eds.), *New Perspectives on Encounter Groups*, San Francisco: Jossey-Bass Inc., 1972. For a highly critical approach see Koch, S., 'An Implicit Image of Man', the same volume.

Section II
The Human Development Concept

2
Towards a Model of Human Growth

Telma Barreiro

1 Importance and need of a theory of human growth or development

Developed society and underdeveloped human beings

The statesmen, economists, and politicians of our age often proclaim their anxiety over the development of nations. Phrases such as 'developed', 'underdeveloped', 'developing', etc. form part of the language of the day, including popular speech.

In contrast, little is heard of the development or growth of the human being. Except among small groups of thinkers concerned about the subject, the problem sounds 'exotic', to put it elegantly. Why does this happen? Why does the well-being of people, their full growth, the quality of their life, never appear as the protagonist of the development process? I think this is probably due to two reasons:

1. On the one hand, it is often assumed that when a society is developed economically and technologically, an adequate development of its members as human beings takes place automatically. In other words, that the material development of a nation is not only necessary but also sufficient for full development of its population. Real, concrete human beings would thus be the natural beneficiaries of this process. In the meantime, however, throughout the process – which in fact is prolonged more or less indefinitely – they are regarded as inert matter, interchangeable and totally malleable parts which can be used as instruments.

    Very few people wonder: 'What will happen with the people if this or that method is applied to incentivate productivity? What will happen with the people if industry is concentrated or dispersed, if this or that form of energy is resorted to etc? What will happen with the inhabitants of this neighborhood if their houses are demolished to construct a complex highway system?' It is taken as self-evident that anything will be good for the common citizen, member of a nation, as long as industries and trade develop, the GNP increases, and the cities are modernized.[1] Our culture is not humanist but, rather, technologistic. One of the proofs of this is the fact that the fundamental centre of studies and academic research officially supported is applied, to a much greater extent, to the technological instead of the human field. The hyper-accelerated development of the machine and automatization and the technological advances which have been attained are dazzling and enormously useful both for the state and private interests. To what extent has this progress helped to really improve human life? How can technological development be directed to provide truly greater well-being to all persons and to improve their possibilities of being happy and becoming fully realized human beings? To what extent has technological growth permitted (or will it permit) a radical improvement of the human species?

These questions sound Utopian and abstract; ambitious men with 'practical' minds (who are, in the end, those who almost always have the political or economic power, or both, and partially the academic and ideological power as well) cannot qualify such 'trivialities' as being important. Furthermore, studies about human beings acquire status and respectability to the extent that they are 'useful' because of their concrete application in industry, offices and workshops, in education, or in the armed forces, permitting an increase in efficiency, productivity, adaptation of individuals to their work, etc.

2 Perhaps another reason why so few people are interested in the subject of the full development of persons as human beings is the assumption that the maximum degree of growth to which humans can aspire has already been reached, the 'pinnacle', so to speak, of their psychic development as a species. Disparaging references are often made towards other forms of culture as being primitive or rudimentary manifestations of the human mind.

It is considered that, thanks to the development of science and technology and the predominance of rationalism and the practical spirit, primitive forms of thought have now been surpassed (such as superstition or animism or polytheism, etc.). Guided by positivism, rationalism, and scientific method, we have been able to construct the civilization and from the mountain top we haughtily regard the deficient, superseded state of people of other cultures, or earlier ages. Our problems of underdevelopment are, in the end, only technical and material, but not mental and moral.

We consider that both the assumptions analyzed above are false. Neither is it true that the rapid economic and technological development of society guarantees in itself and independently of the methods adopted, the full development of the human being, nor is it true that in our culture and our century men and women have attained their maximum growth.[2] On the contrary, we believe that in spite of the immense development achieved in the technological and material plane, in spite of the enormous power obtained for handling and transforming matter and for processing information, the majority of human beings in our culture live in a state of underdevelopment, they are impotent to reach the real unfolding of their potentialities and latent abilities and therefore they are unable to attain inner peace and true well-being. The high rate of suicide, crime, alcoholism, and drug-addiction prevailing in the most developed countries provides strong evidence for this thesis.

The model of growth

Now what do we mean by 'development' and 'underdevelopment' of the human being? Very detailed and subtle scales have been constructed for measuring the economic and technological development of nations, but we are in a state of great poverty when it comes to evaluating or appraising the development of the human being (or of communities, from the point of view of the integral well-being of the population).

Actually, it is not that our society has no model of human development or growth.[3] As in every society, an incorporated model of the 'normal' or 'mature' adult, which is coherent and functional with its economic structure and institutional system, exists in our society. What I want to point out is the following:

(a) In the first place, that model is implicit, it is never analyzed or discussed, nor is any attempt made to support it; it forms part of the social 'fog', of the obvious, of the underlying paradigm which conditions our conduct and our perception of reality without our being aware of it.

(b) In the second place, this growth model is poor and distorting, it does not contemplate the possibility of the emergence and development of aptitudes and potentialities which are fundamental for making a person psychologically rich and potent in his relations with life, with other persons, and with himself.

Statement (a) does not require much theoretical substantiation. It is easy to see that every society has its implicit model of growth or normal development, its concept of a mature person, and also its beliefs regarding which things a person needs to mature psychically and organically.[4] The education of children, socially patterned, aims at promoting in the young member of the community the type of development which is considered desirable. In the light of sociological and anthropological research, it becomes quite evident that the patterns of normal or desirable development of a human being differ from one culture to another and that what each culture will encourage and repress to make its members mature or develop 'normally' will be different.[5]

It is also quite clear that this model is not discussed, that it forms part of the underground layers of social dynamics that are incorporated slowly, firmly, and uncritically by people through the socialization process. This implicit existence of a 'normal' growth model is shown, in our opinion, by the fact that nearly all adults consider themselves capable of educating their children; the majority of adults embark decidedly and unhesitatingly upon the enterprise of forming their children. How could this be explained if it were not by virtue of an internalized model of what normal and desirable human development is, of what a child should be, and what a person should be on reaching adulthood?

Statement (b) is, on the contrary, highly controversial, since it implies a value judgement; and, furthermore, it presupposes another statement also highly controversial: the statement that there exists an 'authentic' or 'ideal' growth, which in fact would need to be demonstrated.

Does authentic growth exist?

Every living organism needs to develop to its maximum, according to its nature. This appears to be a law of life in general. We could think that in the case of a human being this would also be applicable. The humanist philosophers of all times have postulated that the path of morality is that which leads the person to an authentic development of his potentialities and abilities as such.[6]

The problem, however, becomes very complex since human beings always appear to be modelled by culture. Thus it could be asked: Does authentic growth exist, one which is most suitable for human nature? Or are the models proposed by the different cultures only alternative, equally valid models?

The cultural relativist will answer decidedly that they are alternative models, all equally valid. The thesis we maintain here is that there are models much more suitable for human beings, in which their real, objective needs are satisfied more than in others.[7]

In connection with this, four great tasks need to be undertaken:
(a) To make explicit the model of human growth or development prevailing *de facto* in our culture, the model which feeds and is fed by social dynamics.[8]
(b) To draw up a model of authentic growth (and lay its foundations).
(c) To compare the prevailing model or models with the model constructed.
(d) To analyze the structural conditions (economic, social, political, institutional, and psychosocial) which would permit growth in accordance with the proposed model (conditions of feasibility of the model).[9]

My purpose in this chapter is to draw a preliminary outline of a model of human growth; this will be undertaken in part 2. In part 3 I shall deal briefly with psychosocial conditions which promote or hinder human growth according to the proposed model.[10] In part 4 I shall make a comparison between the model proposed and a model implicit in utilitarian culture.[11]

2 Outline of a theory of growth

Primary growth and maturity growth

In the first place it will be useful if, for the sake of analysis, we distinguish between two types of possible growth, which we shall call *primary growth* and *maturity growth* respectively.

Primary growth is usually known and studied as growth or development of the child, the process which occurs in infancy and adolescence and which leads to the final mutation of the child into an adult.

During this stage of growth, there occurs an overt, manifest corporal growth and a rapid psychic development which takes place at the same time and closely connected to the corporal development and organic maturation.

This growth leads the child to the achievement of fundamental skills, such as speech and locomotion, makes possible the development of his thinking from the stage of sensory-motor intelligence up to the acquisition of abstract thought, enables him to learn the fundamental time-space notions and a basic conceptual system allowing him to manage the real world, etc.

Primary growth constitutes (at least in part, or in certain aspects) a biological demand upon the organism. It may become imperfect or deficient to the extent that adequate nutriments are not encountered (both organic and psychic),[12] or to the extent that it is openly impaired by some outside agent or cause (traumatisms, illnesses, intense psychological repression) or even to the extent to which organic maturing is altered by some congenital cause, but in general (and in a first approach to the subject) it may be said that growth in this period of life responds to internal laws of the organism and will take elements from wherever it can in order to accomplish an inner developmental logic.

Within our culture it is generally considered that growth concludes when what we have here called primary growth ends. Human life is viewed as a process in which there is a period of growth (childhood, adolescence), a period of permanence (adulthood) and a period of decadence (old age, senility) – see Fig. 2.1.

The concept of growth is thus defined according to what are considered to be strictly biological limits. 'Natural' limits of growth are

Fig. 2.1 *The process of human life according to the dominant conception in our culture*

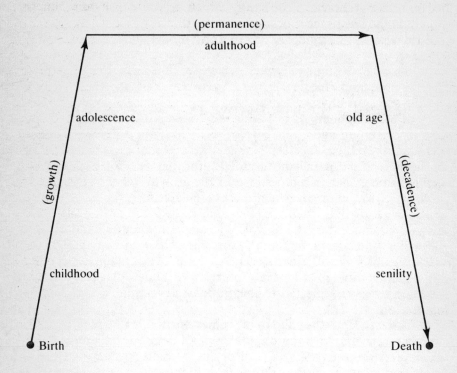

assigned not only to corporal development, of course, but also to mental functions; thus, for example, reference is made to a 'natural' development of intelligence which stops around the second decade of life.[13]

We shall maintain here that in addition to primary growth there exists for a human being the possibility of another type of growth – maturity growth – which does not recognize any limit in time. The intensity and duration of maturity growth is not determined by the individual's genetic code as, at least in certain aspects, physical growth or primary mental growth would appear to be. Rather, it becomes a reality (or is restricted, as the case may be) on the basis of a certain type of interaction with a given environment. Maturity growth is not expressed through an increase in the individual's corporal size or through a rapid, evident acquisition of sensory-motor or intellectual aptitudes; it is slower and less spectacular than primary growth. Neither does it appear to respond, as we have said, to a biological exigency, such as primary growth, but instead to a psychic potency of the human being inserted into a particular type of culture. Hence it is both very difficult to discover and analyze scientifically and very easy to frustrate or restrict because of circumstances connected with the individual's life history and his insertion in the environment. Since it is made possible by psychosocial interaction, maturity growth is a function depending as much on the bio-psychic possibilities of the individual as on the possibilities which the environment has offered him for his personal realization.

While the potentiality for primary growth seems to recognize a biological limit and finds its natural culmination in the attainment of certain stages of physical maturity and certain basic cognitive-affective structures, the potentiality for maturity growth does not recognize, at least *a priori*, an age limit (perhaps with the sole exception of senile regression produced by strictly biochemical processes). On this understanding, the potential growth scheme in the life of a human being would be as shown in Fig. 2.2.

Fig. 2.2 *Ascending line of continuous potential growth*

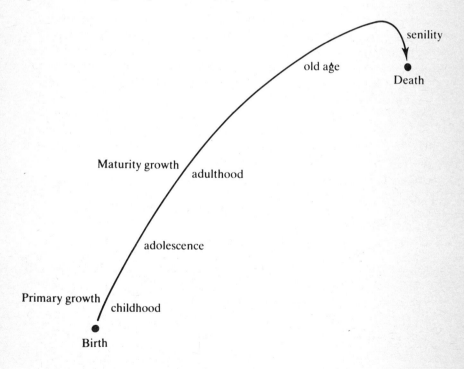

Maturity growth as psychic growth

Maturity growth occurs basically, in the psychic dimension. It is closely linked with human beings' learning capacity and with being open to change and transformations.

Let us pause for a moment over the concept of psychic growth. What do we mean by 'growth' in this dimension? Can one 'grow' psychologically? Obviously, this depends on the use of language and the meaning we attribute to the word 'growth'. In the corporal sense, the idea of growth seems inseparably connected with the idea of increase in physical size. In this limited sense, the expression would not be suitable. Nevertheless, even in the sense of physical growth, the term is by no means reduced to a quantitative increase in corporal dimension because physical growth involves certain transformations which produce qualitative changes too. So it is important to recall that physical growth implies an increase in the individual's potencies and a maturing of his whole organic system. We shall use the term 'growth' in a wide sense which includes but is not limited to physical growth. We shall say that growth consists of development of the psychosomatic power and aptitudes of the individual and, especially in the psychic aspect, of a display of new, more complex possibilities of fruitful interchange between the subject and his environment. In this sense, there is growth of intelligence in the child, which passes from a stage of undifferentiated perception to an unfolding of abstract, logical-conceptual operations. In the development of intelligence there is a constant acquisition of new, more complex, more powerful tools for knowing reality and for an adequate exchange with it, which in some way add to or integrate prior stages of the capacity for adaptation. Each stage thus introduces a perfecting and a complication of the cognitive structure corresponding to the previous stage. Psychologists have designed techniques for measuring degrees of intellectual development and this contributes to make quite natural the notion of growth as applied to intelligence. But it is not only as regards intelligence that the child grows psychologically. He also matures, develops, increases his affectivity, his perception of himself, his grasp, and structuring of social and natural reality, etc. Thus, for example, to the extent that the child passes from a primary, affective symbiosis with his mother to an incipient structuring of the self which leads him to a primary egocentrism and from there to this gradual integration in a group of peers, we can say that he has grown in his adaptation to the environment, in his possibilities for interacting with the world. An extreme example of lack of growing in this dimension is the case of autistic children, closed to communication, blocked in their possibilities of exchange with the world and, for the same reason, with themselves. Maturity growth, which characterized the adult stage of life, is mainly a certain type or expression of psychological growth.[14]

Characteristics of maturity growth

We have said that in our culture it is taken for granted that individual growth reaches its peak in adolescence. We must recognize, however, that in our society it is quite usual to think that a human being often attains in the course of his life an increasing maturity in the sense of an increasingly better adaptation to the environment, a greater mastering of concrete reality. So the importance of accumulated life experience is stressed. Terms like 'mature', 'centred', 'realist', are intended to characterize an adult and differentiate him from a youth or a child.

Our own paradigm of maturity growth differs greatly from what is usually considered to be a process to 'maturity'. The main feature which characterizes individual maturity according to the usual point of view is the capacity to behave successfully within society; instead, our paradigm of growth is concerned about the whole range of aptitudes and potencies of the human being. In our model, maturity growth expresses itself through a strengthening and enriching of personality. It involves a process of construction and consolidation of the self, a development of certain capacities and powers which in adequate conditions, i.e. within a favourable socio-cultural context, should be encountered in a state of active tension or latency in the individual once he has begun to pass through the last stages of primary growth.

Maturity growth reveals itself through the following facets or dimensions of the human being:

1 Profound emotivity.
2 Capacity for communication.
3 Rationality, imaginative capacity, and intuition.
4 Construction of the self and consolidation of personal identity.
5 Sensitivity and experiential openness.
6 Creativity and expressive capacity.
7 Psychosomatic integration (greater security in handling and operating the body).
8 Adult realism.
9 Capacity for constructive work.
10 Impulse towards active social transcendency.

We can say that maturity growth in the individual exists or occurs in some of these aspects in the following cases:

(a) There is a consolidation and increase in the capacity to establish authentic profound affective relations with other individuals, the capacity to offer adult love, the capacity for surrender, and enjoyment in communication; capacity for empathy increases and childish egocentrism recedes. (This implies growth in the dimensions we have called 'capacity for communication', profound emotivity', and 'impulse towards active social transcendency'.)

(b) There is consolidation of rational capacity, capacity for integrating knowledge in ever more encompassing systems, and capacity for understanding reality; at the same time there is development of imagination, cognitive curiosity, capacity for wondering. (This implies growth in the dimensions we call 'rationality' and 'creativity'.)

(c) There is a progressive surmounting of the affective dependency bonds (of a symbiotic nature) with the primary parental group and other substitute forms of entities; (real) security in own identity ('ontological' security)[15] is acquired and advance is achieved in the construction of a real or authentic self,[16] abandoning increasingly artificial masks used for the social presentation of the self.[17] Fear of and submission to authority (and other irrational fears)[18] are overcome and a gradual moving away from external, formal rules is achieved; there is progress in self-knowledge and in introspective power; a greater coherence and unity in personality are achieved. (This implies growth in the dimension we have called 'construction of the self and consolidation of personal identity'.)

(d) There is an opening and deepening of the capacity for aesthetic and sensorial pleasure and for integral erotic pleasure. (This implies growth in the dimensions we have called 'sensitivity and experiential openness' and 'psychosomatic integration'.)

(e) Creative potentiality develops, unfolds, and is consolidated, and new, fruitful ways of personal expression are found (growth in the dimension 'creativity and expressive capacity'.)

(f) The harmonious integration of the adult's own image with personal indentity is consolidated; there is an unfolding of corporal powers (as regards movement, elasticity, sensory-motive co-ordination, strength, etc.) and of the accompanying pleasure in the use of these powers. A more satisfactory and integral operation of the body is achieved (growth in 'psychosomatic integration' and in 'expressivity' at corporal level).

(g) There is an increase in the capacity of understanding and (creative, critical) adaptation to reality, in the capacity to face and solve conflicts at more and more satisfactory levels of integration and equilibrium; the mechanisms of regressive escape recede (search for recovery of 'uterine equilibrium' growth in the dimension 'adult realism').

(h) There is consolidation of the capacity for constructive work with personal motivation and creativity (non-mercenary, non-bureaucratic, non-prostituted) and with a sense of social transcendency. Perception of the 'other person' becomes more acute, and the sentiment of solidarity and zeal for justice become stronger (growth in the dimension 'impulse towards an active social transcendency').

3 Is maturity growth possible? (Conditions which hamper and conditions which foster maturity growth)

Maturity growth is rooted in primary growth

In part 2 we established the difference between primary and maturity growth. The difference enables us to see maturity growth as a growth which occurs in the innermost recesses of the individual's psychic life, distinguishing it from growth of a somewhat more 'natural' or 'biological' nature and emphasizing that it can only happen after certain basic stages of psychic life have been attained.

It is now time, however, to point out that such a clear-cut distinction between the two types of growth does not exist and that maturity growth sinks its roots deeply into the primary growth stage. It also becomes necessary to ask whether primary growth is really universal and identical in all cultures and even in all individuals from the same culture and whether this growth is accomplished 'innocuously' or in an axiologically neutral way. Quite the contrary, the way in which primary growth develops is to a great extent conditioned culturally. Furthermore, the way in which primary growth develops in an individual will heavily condition (either negatively or positively) the possibility of that person's subsequent maturity growth.

In the individual's life history the first significant events are those which concern the satisfaction of basic biological and psychic demands. As mentioned before, it has been demonstrated that, at the biological level, deficient nourishment during the first years of life produces a cerebral underdevelopment which is irreversible and will obviously hinder all subsequent integral development. It is therefore clear that as a necessary condition to attain maturity growth, the individual must be assured in childhood of a basic level of organic well being. While this is a necessary condition, it is far from being a sufficient condition. There are many psychic needs which should be satisfied if maturity growth is to be fostered or encouraged from earliest childhood. The period we have called primary growth is strongly determined by the influence which 'significant adults' exercise on the moulding of the child's personality. In this first stage of life the possibility for developing or blocking later maturity growth is already being defined. Thus, for example, children must have affection, recognition, and active protection from their primary parental group. They must be stimulated and encouraged in their aptitudes and potentialities. They should not be overwhelmed by the weight of irrational authority.[19] They should be given adequate sensorial, sensory-motive, intellectual, and affective encouragements which foster or stimulate the different stages of their development. They should be able to join in peer groups where solidarity, mutual respect, and self-expression are encouraged. Their curiosity and eagerness to investigate should meet with an attitude of approval and stimulation. These would be some of the ideal or optimum conditions which would pave the way for integral maturity growth.

This is but a compressed list of the variables which come into play in the primary growth process and which affect the subsequent maturity growth. To understand properly the nature of the link between one stage of growth and the other, it would be necessary to develop in detail at least some of these aspects to see how they positively or adversely affect the individual's growth in the different dimensions pointed out in the final section of part 2.

Let us take just a few examples as illustration. A child who passes his first years of life in a primary parental group where he is systematically treated with contempt or disdain, where he finds no basic affective support or recognition, or acceptance of his person, will have little probability of developing a positive self-image which would allow him to attain a suitable construction of his self and his personal identity. Since this latter is a necessary condition for reaching an adequate level of communication and surrender in human relations, this person will be handicapped for achieving adequate maturity growth in several of the indicated dimensions (e.g. in 2, capacity for communication, and 4, construction of the self; and very probably in 6, creativity and expressive capacity, and 7, psychosomatic integration).

Another example could be one referring to rationality. Development of rational capacity begins in earliest infancy and is the result of a slow, complex process. If during the education of a child, emphasis is given to shibboleths and dogmatic rules of conduct imposed in an authoritarian manner, if he receives few sensorial, affective, motive, verbal, and intellectual stimuli, if his educational training is limited, for example, to passive mechanical, coercive intake of data, all this will have a negative effect on the growth of rational capacity. His cognitive structures will probably remain at a level of development which will prevent him from attaining full maturity of rationality.

Moreover, there are also subtle links which interconnect variables from different fields. For example, up to what point can a weak self-perception retard development of intellectual capacity? If we were to explore in what way the different dimensions of maturity growth are connected with aspects of primary growth and are subordinate to it, we would have to go deeply into psychological theory and to follow (no doubt critically) the tracks explored by current developmental psychology.

Maturity growth and culture: psychosocial conditioning

The way in which the individual passes through the stage of primary development is not the only thing which affects the nature of his subsequent development. The child slowly widens the scope of his contacts with his environment and integrates into overall society. Culture, which in childhood is conditioned and transmitted to the child only by his primary groups, begins to operate more directly on the individual and determines the ways and patterns of his psychological evolution, the mechanisms and facets of his psychic life which can be encouraged and those which must be repressed. There are many psychosocial mechanisms to which the individual will have to adapt himself, which form an unavoidable part of his daily life, which contribute to modelling his character, the type of his aspirations and desires, his expectations, as regards his own life and personal evolution. In fact, the general socio-cultural environment is so determinative, it conditions the development of human life to such an extent that there is not much sense in designing a model of human growth without studying closely at the same time which are the psycho-social conditions which make it possible for growth to take the desired direction and which are the ones which restrain, hamper, or deflect it. This means analyzing critically a certain social reality and proposing 'ideal' models of social functioning. A task like this implies an ambitious work project. We shall limit ourselves here to pointing our some lines which we consider significant and important to explore.

1 The adult and his work
We believe that in the adult's world there are some 'key' elements which have to do with the promotion or mutilation of integral growth. In the first place, his relation to his work. It is difficult for an adult to achieve integral growth if he is conditioned and forced by circumstances to have an alienating kind of work. For an adult, his work is the axis of his life. This is so at least for three reasons: First, because work fills a great part of the waking hours of his life. Second, because through his work he attends to the basic survival needs of himself and his family; the possibility of losing his job or having conflicts in his work generates an enormous anxiety or tension for all that it means in individual life and, above all, in family life. The third reason why work occupies such an important place in an adult's life is that the opinion or social evaluation made of him will depend heavily upon his kind of job.

For these reasons the psychosocial conditions in which a person develops his work have a tremendous impact on the way in which his development will take place. For maturity growth to be possible, the individual should have the possibility of becoming related maturely and creatively to his work, he should have the possibility of projecting himself, of transcending in his work, of feeling his daily labour as a form of realization and not, as occurs in the majority of cases within societies such as ours, as a painful, oppressive obligation. We can say that the circumstances which surround the world of work within a utilitarian society are generally far from being suitable for promoting integral development of persons. Some of these circumstances are:

(a) The worker does not participate in the organization or planning activities of his work. He carries out only the small part which has been assigned to him without knowing the whole nor having participated in its planning.
(b) He is submitted to an authority which orders and directs him and which he must obey; he must follow instructions without understanding or questioning what they mean.
(c) He must comply with rigid timetable and production regulations; his task is often monotonous and hard; there is rarely any opportunity to apply his creativity or ingenuity; he is rarely asked to give an opinion.
(d) He has a 'mercenary' relation with his work; he works for payment but does not enjoy the product nor can he find pleasure in his work (partly because of conditions 1, 2 and 3).
(e) With the inducement of a promotion, the worker is sometimes led to compete with his own companions.

Going over these characteristics, we see that the worker cannot behave really like an adult person in his work; rather, he is obliged to abide by instructions and respect authority like a well-behaved child in order to get his reward (the wages in the case of work; affection and recognition in the case of the child).

Moreover, apart from these characteristics which are more or less general, on many occasions the very nature of the work requires great physical effort and/or psychological violence, which cause serious impairment to the person (this would be the case with work involving intense noise, release of noxious gases or substances, very high or very low temperatures, and involving the transport or manipulation of very heavy objects, etc.).

If the psychosocial conditions were favourable, the link with his work would conduct the person towards growth in the dimension we have called 'adult realism' (8 in the list in part 2), 'capacity for constructive work' 9, 'impulse towards active social transcendency' 10, and 'creativity and expressive capacity' 6. At the same time, it should reinforce other dimensions, such as rationality, communication capacity, and consolidation of personal identity.

However, given the circumstances prevailing in utilitarian societies, it is difficult for this to occur. On the contrary, adaptation to the world of work is one of the most frequent sources of frustration in integral maturity or full growth.

The usually accepted model of maturity is that of a man who can accept and endure with 'fortitude' an 'adequate' relation with his work, a relation which is in fact, because of the circumstances already pointed out, a source of personal mutilation.

2 The adult and public life
Another aspect to be considered is participation of the individual in the handling of public affairs. If maturity growth is to be encouraged, the

individual must have the possibility of really participating in the groups which make decisions that affect the direction and organization of social life. Real, effective participation in decision-making which affects the course of the social world around him and in which he is immersed would give the individual the possibility of feeling his power, capable of having an effect on the destiny of his own life and that of his community, and this would promote his maturity growth in different dimensions. But if an individual is obliged to obey and remain as a mere passive observer of decisions which affect society, and therefore his life, his family and his neighbours, he cannot truly reach a full integral degree of growth. In this case it is our opinion that the maturing of the same facets as in the previous case is thwarted.

3 The adult and human relations
Another fundamental variable, which cuts across the other two in a dynamic interaction, is that of the prevailing model of human relations. This model within a society is generally functional or adapted to the exigency of the society in such a way that the individuals may accomplish 'naturally' the roles and take part in necessary forms of interaction which maintain the social institutions. This concept is linked with what Fromm has described as 'social character'.[20] The model of human relations is generally incorporated at an early stage in the child and then reinforced through his interaction with the environment.

The type of human relations which the individual incorporates into the basic structure of his personality and his motivational system will have a strong influence on the scope of dimensions 1 (profound emotivity), 2 (communication capacity), and 5 (sensitivity and experiential openness). It seems fairly obvious that if an individual is socialized within a model of very good human communication, of generosity and sincerity, in a model where sensitivity and affective openness are considered valuable, his experiential and emotional worlds will become enriched and his communication with others will be open and fluid. If, on the other hand, the model of human relations which arises from the institutional play and its structural constraints leads to the isolation of the individual and to a defensive-offensive attitude, the person will be condemned to psychological under-development. If the prevailing ideal is one of competition, of individual competitive success, it is obvious that it is contradictory to alleged sentiments of solidarity or human love. Thus, for example, the relations between the customer and the merchant, between the buyer and the seller, the boss and the worker, existing in a utilitarian type of society, are essentially antagonistic relationships which cannot generate (at least systematically) attitudes of affection and mutual support or help.

In this case growth is limited in the dimensions 1 (profound emotivity), 2 (communication capacity) and 5 (sensitivity and experiential openness), while even the development of 4 (construction of the self and consolidation of personal identity) and of 6 (creativity and expressive capacity) become difficult, to the extent that the individual must protect and hide himself instead of showing himself and expecting from the other affective and sympathetic approval and a positive support for consolidation of his self-esteem.

4 Importance of groups for personal growth process; the 'healthy' groups

Since a large part of the process of socialization and incorporation of culture as well as a large part of the process of construction and consolidation of the personal self become possible through the individual belonging to small human groups, we can categorically state that the groups to which the individual belongs throughout his life play a decisive role in the nature of his development. The different ways in which an individual's psychological development can follow depend decisively on the small groups to which he has belonged.

Small groups reflect of course the features of their overall society. It is very difficult to make a social 'island', at the margin of the rest of the psychosocial mechanisms which dominate a culture. It would therefore not seem to make much sense to analyze a group as if it could generate or mould itself independently of the whole institutional inter-play of the society. In this sense, the study of groups is not unrelated to the study of the aspects we have just introduced in paragraphs 1, 2, and 3, since the human groups are inserted in institutional mechanisms and receive from them their influence and to a certain degree their character.[21] Nevertheless, we believe it important to pose the subject as significant on its own for our problem, because direct human contact, loaded with affectivity and implicit messages, which characterize small groups (whether in the family, the factory, the office, the neighbourhood, the school, etc.), has a direct, particular influence on the structure of personality. Even when culture and overall society penetrate and dominate the general vision of the members of a given society, the development of individuals is to a large extent a dependent function of the groups to which they have really and personally belonged. It is not something mechanical which derives automatically from the global social structure. It is a process involving persons who, while they reflect in some way the social character, also have individual characteristics.[22]

Thus we can state that if a person is to reach full psychic development it is necessary that he has the possibility of integrating into one or more human groups which encourage or promote this development. It is very difficult to be in the process of continuous maturity growth in solitude. Maturity growth requires the stimulus and support of other human beings.

Not every small human group, however, encourages or promotes the integral growth of its members. On the contrary, there are certain groups which by their characteristics are more likely to atrophy or mutilate this process (partially or totally).

We shall say that a human group is healthy when it promotes or encourages the maturity growth of all its members, and that it is sick when it tends to atrophy it. In general, groups usually have both healthy and sick elements or aspects. A healthy group possesses certain fundamental features with regard to the nature of leadership, the type of connection established with its members, the type of expectation which it encourages regarding other human groups, etc.[23]

If a person were fortunate enough to be able to join a group with these characteristics, he would reach a certain degree of basic security and would be encouraged to express himself freely, to release his creative powers, and to transcend positively towards the group. It is most certainly true, however, that in a culture where individualist, competitive, and utilitarian attitudes predominate, there is little likelihood of integrating groups which function in accordance with these patterns. Even within the family nucleus where it is supposed that the individual will find his principal affective refuge, some of these conditions are often absent. The adult individual is inclined to place his unconscious hopes of a healthy group in the couple, where a relationship of affective symbiosis often occurs, perhaps by the effect of contrast as regards the outside environment. But symbiosis is a stumbling block in the way of personal growth.

5 The material conditions of life

The material conditions of life must meet fundamental basic requirements if a person is to develop in a complete, adequate manner.

Strictly speaking, the maturity growth model may be inapplicable and even a cruel irony if it is attempted to refer it to human groups whose basic fundamental survival conditions have not been satisfied.

The possibilities of achieving integral, harmonious development will be in fact drastically and dramatically reduced for any person who is undernourished or afflicted by endemic diseases characteristic of poverty, who must dwell in precarious, sub-human conditions, who is obliged to carry out brutalizing work. In this sense the material conditions of life are an unavoidable condition which, if not suitably covered, will cruelly retard maturity growth by affecting in the first instance primary growth.[24]

Therefore, a fundamental requirement for dignifying human life is to eliminate social, economic, and political conditions which maintain a large part of the world's population in a state of physical and psychical misery.

The dilemma which arises here is whether the steps to be taken to solve this first priority problem of mankind should face from the very beginning the issue of psychic growth or if it would be better to postpone it for more advanced stages of the social change process. In our opinion, both the analysis of social change and the design of different models of societies must indeed introduce the problem of psychic growth from the start. From this standpoint, authoritarian and paternalistic policies of social transformation have to be rejected.

Now, just as we are convinced that a minimum of well-being and material or biological security is a necessary condition, we are also sure that it is not a sufficient condition. It is not only the bad material conditions of life that limit the possibility of maturity growth. A person can have all his basic needs satisfied and even more, satisfied in excess of his biological needs, but he can be essentially at a standstill in his psychological development. In this sense, he may present the apparent paradox of a socially powerful man who is weak and impotent as regards his own personal growth.

The fact is that it is not only socio-economic factors which limit and condition maturity growth but also cultural patterns, the system of values and the different cogs in the psychosocial mechanism, such as those we have been analyzing in the previous paragraphs.[25] In our view, therefore, it is not arbitrary or frivolous to introduce this subject when referring to possible ways of overcoming the economic underdevelopment of a nation, because the different styles of social development carry with them different patterns of human growth.

4 Presentation of two paradigms: success versus maturity growth

Utilitarian culture as opposed to maturity growth

It follows from part 3 that the psychosocial conditions which characterize a utilitarian, individualist, and competitive society are not the most favourable for encouraging maturity growth.

We could take as a significant indicator of the social expectations in this sense the stereotyped figure of the successful man within a culture of this type. We shall be able to see that a man of 'success' is not precisely one who has experienced a true maturity growth.

The successful man in a society of this kind is one who has amassed fortune, power, prestige or fame (whether in the sphere of business, politics, intellect, sports, etc.), or a combination of some of these attributes. In general (but not necessarily always), the man who devotes his life to the achievement of these goals, is one who must neglect his inner self, whose interior growth is forsaken.

As a paradigmatic example we can take the prosperous businessman, who has been able to acquire a fortune, who provides his family with an abundance of goods and solid comfort, who enjoys prestige and admiration and arouses considerable envy. His main concerns revolve around his business. He must maintain his prestige and outward composure, his image of a determined, 'aggressive' man. The greater part of his time has to be devoted to paying attention to his possessions and his business affairs, under penalty of losing the position achieved. Since he had to concentrate a large part of his psychological activity on the effort of making his way in the world, which has meant overcoming many difficulties, this man was probably condemned to forget about himself, to amputate certain dimensions of his psychism which constituted obstacles to adapting well to the environment.

Now this is one of the alienating mechanisms of the self: the mutilation of potential psychism, which hinders and atrophies maturity growth. The other important way for 'good' adaptation is to create 'correct' defensive-offensive mechanisms of the self, to raise solid barriers to avoid being battered by the medium, to avoid being destroyed by external aggression. This leads to the armour-plating of the self: when the individual feels compelled to act in an essentially aggressive medium, within which his vital option is to adapt himself and win or not adapt himself and fail, he becomes obliged to create impenetrable barriers through which outside aggression cannot infiltrate. This enables him to offer an image of security and power that discourages his supposed adversaries and disguises his primordial need of affection, protection, and recognition from his fellow beings.

Armour-plating of the self almost always reinforces the mutilation of potential psychism. Constant camouflaging of the inner tendencies, the dissembling of himself which the man is compelled to do in front of others to avoid being taken unawares, leads to a profound self-convincing that the image he presents to others is his true image. After many years of struggling against himself, this man will have managed to assimilate himself to the dominating social model at the price of having atrophied his individual originality and his potential experiential wealth. We can say that it is the defensive offensive mechanisms of the self which the individual is led to construct form a castrating armour which puts a limit on the possibilities of fundamental communication with others and the creative search of his own individual possibilities.

Two paradigms: success versus growth

Within the prevailing view of what constitutes a successful man, we can point out a series of typical achievements. It would be interesting to compare these achievements or acquisitions which are characteristic of the successful man of utilitarian culture to the achievements which could be attained by the mature or developed person who lives out a real, profound process of maturity growth.

This comparison will be made using ideal or pure types, emphasizing the more significant features. We shall divide the description of the paradigms according to the different environments or aspects which come into play in different achievements (e.g. 'human relations', 'power', etc.). It will be interesting to observe the character of instrumentalization which behaviour adopts in the first paradigm (subordination of achievements to the ultimate purpose of obtaining power, prestige, social status, etc.), while in the second paradigm each achievement has its own value, in itself profoundly satisfying for the individual.

Table 2.1 *Comparison of two paradigms (ideal types)*

Area	Achievements of the successful person, well adapted to utilitarian culture	Achievements of the person who has achieved maturity and strength through his maturity growth
The family	To have a worthy, respectable family, a spouse with good socioeconomic and cultural status, well-adapted children who will be future successful citizens.	To construct and maintain profound, mutually gratifying, lasting family ties, which allow a real, independent, and personal development of each family member.
Human relations	To enjoy social prestige; to be admired and envied for one's social status.	To be capable of feeling positive sentiments of love and affection towards other human beings and of arousing these sentiments in others.
Power; social groups	To occupy positions of power; to know how to command; to be able to dominate others firmly and with authority.	To integrate in healthy human groups. To have no desire to dominate or be dominated; to have no yearning for power. Not to depend unduly on the opinion of others.
Social relations	To maintain social relations of a suitable level, convenient for one's own social promotion.	To communicate with other individuals in depth and without a defensive/offensive mask; to try to understand and value others and create a bond of mutual support with them to enjoy friendship for its own sake.
Work	To have some kind of activity which confers prestige, security, and fortune. To instrumentalize one's working capacity adequately to meet these objectives.	To be absorbed in a personal constructive work, directed at making something of one's own creation.
Adaptation and realism	To be able to insert oneself into social reality with great realism and capacity for adaptation (instrumental	To insert oneself into social reality with a critical, non-mimical realism in order to transform it and improve it.

Table 2.1 *Continued*

Area	*Achievements of the successful person, well adapted to utilitarian culture*	*Achievements of the person who has achieved maturity and strength through his maturity growth*
	mimicry) in order to be successful, outstanding.	To aim at social transcendency and the welfare of the community.
Reason, knowledge and research	To understand reality in order to adapt to it; to have original ideas, to discover, to investigate, or invent for winning fame, prominence, success. To create objects, works of art, new techniques, etc., for achieving success, prestige, or money.	To think, to meditate creatively and critically, for the intrinsic pleasure of doing so and for the contribution it may mean to the common good. To have creative power, to be able to mould or rebuild aspects of reality, to improve it, to express the inner abundance and bestow it on others.
Self-knowledge	To know oneself well enough to act conveniently, to control oneself and project the best possible image.	To know oneself as profoundly as possible in order to acquire more coherence, greater psychological openness and capacity for communicating, and a richer psychic life.
Sensitivity	To control feelings and emotions adequately so that they suit the medium and to prevent being hurt. To dissimulate and hide emotions which would make one vulnerable. To postpone emotions which complicate the struggle to succeed. To instrumentalize adequately the 'aggression-defence' mechanism and the functional concealment of one's true self.	To experience feelings and emotions intensely. To keep alive primary sensitivity but enriched and mature, spiritually dense, laden with constructed personal values. To feel empathy and enjoy affective surrender.
The body	To use one's body adequately so as to be successful in life; to enjoy good health for working well; to have a strong, well-formed, athletic, elegant body which helps to make a good image. Perhaps to be champion in some sport and thus win fame, prestige, and money. To enjoy physical pleasures. Sexuality lived as a separate domain, unconnected with sensitivity, vitality, communication.	To feel one's body as fully integrated with one's psychic life and as a vehicle for communication. To enjoy movement, strength, ability heedless of success or rivalry with others – only as an expression of vitality and one's own potentialities. To link erotic enjoyment with communication and affective surrender and with the sheer thrill of vitality.

Table 2.1 *Continued*

Area	Achievements of the successful person, well adapted to utilitarian culture	Achievements of the person who has achieved maturity and strength through his maturity growth
Learning, beliefs, and changes	To learn enough for work and social adaptation. To change sufficiently for adapting to an objectively changing situation. To avoid becoming passé or out of date. To know what a person of a certain status supposedly ought to know. To be able to convince others, by instilling into them ideas or values which suit one's purpose.	To have capacity and readiness for change, in order to transform oneself, to learn from experience to incorporate new ideas and emotions, integrating them into one's personal system and possibly reconstructing the system. To maintain psychic malleability. To foster the same malleability and capacity for dynamic learning in others without attempting to impose dogmatisms or rigid beliefs. To acquire abilities and knowledge which enable one to integrate one's personal system in a progressively harmonious manner and to intervene more efficaciously on behalf of the medium and of one's own search for personal realization.

An antinomy of utilitarian culture: Individualism versus vacuous individual life

We have attempted to show that utilitarian culture does not supply, at least at the level of its more generalized mechanisms, stimuli for maturity growth.

Moreover, utilitarian culture generates an irreducible contradiction: while it urges individualism, while it converts individualism into an ethical-anthropological postulate and stimulates the tendency of the individual towards his personal success, it offers as a model of individual realization a paradigm sadly lacking in psychological achievements, limiting on general lines the success of individual life to the obtaining of satisfactions such as economic well-being, social prestige, or power. These achievements, which are pursued as ultimate objectives or goals, do not lead to the increase of individual psychological richness or power but contrariwise, as we mentioned earlier, in the struggle to obtain them, the individual impoverishes or dries up the source of his potential psychological growth. Therefore, a culture of this kind tends to create a state of constant anxiety and insatiability in its members. There is a sort of tension or basic anxiety for individual plenitude which is constantly diverted, within a culture such as ours, towards the possession of more and better material goods. This appetite will never be satisfied, because the real need is connected with the capacity or psychological power of the individual to live, to fully experience, to communicate with others, to love other persons intensely and be loved with the same intensity, to thrill with emotions and sensations, to transcend towards others, and to be projected into a creative activity and constructive work within a community, which enables him to be projected beyond his own fleeting existence.

Many people attribute their disappointment with life to the fact of having been unable to achieve all the comfort, security, or material economic well-being they wish. This mask of material dissatisfaction nevertheless hides a much more radical dissatisfaction: it hides the void of a spiritually mutilated, atrophied life, limited at an early stage in the dynamics of its development.

Notes

1 Perhaps what really happens is that the leading protagonists of this process are the economic and political interests which come into play; to the extent that each economic or political group seeks its own advantage, individual people are only useful pawns in this game of power and economic benefit. It could therefore be thought that it is the State which takes care of the well-being of people (the thesis of the Welfare State) but this is also relative, since each government in turn looks for its own objectives, which are related to the consolidation of power, supremacy over other nations, etc.

2 It is essential to clarify that we do not under any circumstances deny the decisive importance of material conditions for a decorous human existence and for full growth of a person. We consider that the satisfaction of the so-called basic needs of all the inhabitants of the world is an absolutely primordial urgent condition, a challenge to the moral conscience of all human beings who enjoy life without fundamental privations. Moreover, discussion which centres around the question of human development but does not contemplate the incidence of material conditions of life and everyday experience upon people, appears to me unacceptable.

My point of view regarding the connection between economic development and human growth can be broken down into two aspects: (i) Referring to the so-called developed countries, where the basic needs are covered for the great majority or the whole of the population and where highly sophisticated industries, advanced technologies, etc. have been developed. (ii) Referring to the so-called 'developing' countries, where an important part of the population suffers penury, malnutrition, endemic diseases, illiteracy, promiscuity, etc.

As regards (i), the question lies in analyzing up to what point technological development has really helped human growth and up to what point it has atrophied or deflected it. Here it is necessary to study what were the real objectives and methods with which the material, industrial, and technological development took place and what results did this development model have on the profound psychic well-being of the population.

As regards the countries where problems of poverty subsist on a massive scale, the point is to examine what will be the objectives and methods to encourage the promotion of their economic-technological development since – as demonstrated by the developed countries – this will condition the whole form and quality of life of their inhabitants.

The controversy arises from three different attitudes: (a) It is essential to obtain economic development by any means whatever, because it is a necessary and sufficient condition for attaining well-being, as has been proved by the developed countries. (b) It is necessary to obtain economic development by any means whatever. We do not know if that is a sufficient condition, maybe not, but that is not the important point. What is really important is that economic development is a necessary condition for raising millions of human beings out of misery. The methods do not matter, perhaps it is necessary to sacrifice some generations; the final objectives are what matter (finally well-being will be achieved). (c) It is necessary to attain a form of economic-political-social transformation which, while radically satisfying the basic needs of all the population, lays the foundations for the development of a healthy society which can integrally encourage the growth of its members. The methods do matter because they decisively condition the ends. Furthermore, it is not licit to sacrifice one generation in favour of another. This final attitude is the one which is assumed here.

3 When referring to 'our society', I mean the type of society which characterizes the developed Western countries, sometimes known as 'central' or 'Northern countries'. In other words, it is the type of society which can be briefly characterized by its competitive individualism, its high evaluation of political and economic liberalism and of technological development, by consumerism, and the use of publicity in the service of consumerism, etc. Many of the characteristic traits of these societies are shared, moreover, by the developed, socialist-economy countries, to the extent that it would be incorrect to consider them as exclusive to capitalist countries. As regards Argentina, while it is not considered a developed country from the economic viewpoint, its dominant socio-cultural and economic model is strongly imbued with the Western, capitalist-development model described above. I refer later on to the dominant culture within this type of society as 'utilitarian culture'. (See Alvin Gouldner, *The Coming Crisis of Western Sociology*, New York: Basic Books Inc., 1970, first part, paragraph 3).

4 The variety of the organization of the different motivational systems in the different cultures is truly remarkable. By way of example (more picturesque than scientific) we quote below two texts: one describes the expectations and desires of an average North American of the present day, as perceived by Rosser Reeves, a first-class salesman whose work consists precisely of knowing what people want. The other text is taken from the *Iliad* and shows the motivations and ideals of a great warrior of heroic Troy. Rosser Reeves says: 'We know, for example, that we don't want to be

fat. We don't want to smell. We want to have children who will have good health and we also want to enjoy good health. We want to have beautiful teeth. We want to dress well. We want people to like us. We don't want to be ugly. We are looking for love and affection. We want money. We like comfort. We hope to have a nicer house. We want honesty, self-respect, and a place in the community. We want to own things that will make us feel proud. We want to be successful in our work. We want to enjoy security in old age.' (From G. Miller, *Psychology: The Science of Mental Life*, New York: 1966.)

In *The Iliad*, when Hector is about to leave for the war, his wife Andromache tries to hold him back: ' "Oh Hector! Your bravery will be your undoing. You have no pity for the little child nor for me, unhappy me, who will soon be your widow. Because the Achaeans will all assault you and destroy you. It would be better that the earth swallow me up if I should lose you because if you die, there will be no consolation for me, only sorrows . . . So have pity on me, stay here in the castle. Don't make the child an orphan and your wife a widow." The great Hector, his helmet plumes waving, replies: "All this worries me, my dear, but how I would blush before the Trojans and the Trojan women in their long dresses if I were to flee from the battle like a coward. Nor does my heart prompt me to do so, I always knew how to be valiant and fight in the front line among the Trojans continuing my father's immense glory and my own. I know this very well and I know it is certain that one day the city of Troy and Priam and the people of the wealthy Priam will perish. But the future misfortune of the Trojans, of Hecuba herself and King Priam . . . do not worry me as much as what you will suffer when one of the Bronze-armoured Aquaeans carries you away weeping, taking you captive. And perhaps someone will exclaim, on seeing you weep: That was the wife of Hector, the greatest warrior among the Trojans the horse-breakers, when they were fighting around Ilion . . ." And so saying the illustrious Hector stretched out his arms to his son: "Zeus and all ye gods! Grant this child of mine be, like me, illustrious among the Trojans and just as courageous; may he reign powerfully in Ilion; may they say of him when he returns from battle: He is more valiant than his father, and that when laden with the bloody remains of the enemy he has killed, he will make his mother's heart rejoice." '

5 See, for example, the interesting testimony of a maturity concept quoted by Fromm in *The Sane Society*: 'I define maturity, says Dr Strecher, 'as the capacity to persevere in a job, the capacity to yield in an occupation more than is requested of one, truthfulness, persistence in carrying out a plan to the end in spite of the difficulties, capacity for working with other persons within an organized group and under an authority, capacity for taking decisions, the will to live, flexibility, independence, and tolerance.'
6 Regarding the tradition of humanistic ethics in this direction, see among others, E. Fromm, *Man for Himself*, New York: Ballantine Books, 1965, chap. 2.
7 With regard to what I understand by objective necessity, see my article 'Does It Make Sense to Investigate Human Needs?' presented at the Second Latin American Meeting on Research and Human Needs, Montevideo, June 1978.
8 By defining explicitly the prevailing model, the possibility of maintaining

double standards and axiological incoherence is reduced. Many times, at the level of declarations, manifestos, or public speeches, an image of the human being is postulated which has nothing to do with the true prevailing model. According to this, there would be something resembling a double standard: an alleged morality and a real morality. On making the real model explicit it would be given a public, controversial character, allowing analysis and polemics, weakening hypocrisy and self-delusion.

9 Our thesis is that a model of ideal human growth underlies not only every society but also the social and human sciences. Some sociological currents, for example, are conformist and naturally adopt the dominating cultural patterns as part of the social fog which also envelops the researcher and his internalized system of values. Something similar occurs with some psychotherapeutic and pedagogical currents. To unfold a desirable human growth model analytically would provide certain epistemological benefits, to the extent that this would permit:
   (a) Analysing the theory drawn up, from the point of view of its internal coherence (logical analysis).
   (b) Attempting to demonstrate it (or refute it) theoretically (theoretical integration attempt).
   (c) Attempting to prove it (or refute it) empirically (empirical contrast attempt).
   (d) Confronting it with alternative theories and comparing them both from the theoretical and empirical point of view and from their internal coherence.
   (e) Analysing whether this model of growth exists in our culture or some other and if so, which.
   (f) Analysing the theories of economic development of the nations in the light of the theory of development of human beings (for example, to what extent does the strategy of economic and technological development followed by the so-called developed countries contribute to promoting growth as we have defined it?).
   (g) To confront the different psychological, psychotherapeutical, and educational theories with this theory of growth.
   (h) To study psychosocial factors which may arrest this growth.
   (i) To investigate which are the traits or characteristics of a society which would promote such growth.
   (j) To evaluate any given historical, real culture in the light of this theory.
   (k) To analyze the problem of social sickness (drugs, suicides, murders, sexual violence, alcoholism, etc.) in the light of this theory.

A theory of growth offers a theoretical framework, an explicit conceptual paradigm within whose limits a psychological, anthropological, educational, or sociological hypothesis can be evaluated, integrated, or rejected.

10 One very important point which is not broached in this article is the theoretical and empirical foundation of the proposed model. I am aware that this undoubtedly weakens the presentation of the model and makes it vulnerable. It would also be necessary to deploy each of the facets or dimensions of growth and their mutual interaction with much greater detail so that this enumeration is not something abstract and difficult to visualize

in concrete human reality. Another aspect which would require deeper study concerns the types of social structure which encourage or stunt growth according to the model presented. This aspect is crucial for a defence of the model since it would remove it from the merely ideal plane on analyzing its structural and psychosocial conditions of feasibility. All these are in fact different steps of a theoretical work on which I am engaged.

11 See Note 3.
12 The enormous importance both of organic nutrients and of psychological stimuli for the normal development of the organism has been demonstrated by several recent researchers in biology and psychology (for example, the studies by Dobbing and Sands in Manchester, Mark Rosenzweig in Berkeley, etc.). One of the findings was that during the first four years of life the lack of adequate nutrition produces a deterioration in brain development by altering the normal development of the connective nerve tissue and myelinization. According to these studies, children who have been deficiently nourished during their first years of life will suffer an irreversible underdevelopment of the brain which will obviously affect their intellectual yield. With regard to sensorial stimuli, it was possible to prove in experiments with rats that the presence of richer sensorial stimuli fosters a greater development of the cortical zones of the brain. Work has also been done on this subject in psycho-pedagogic experiments known as 'compensating programmes' in which it was attempted to provide better educational opportunities to children from a culture which gave them few stimuli and development possibilities. The changes in the levels of development achieved were significant. (See: Salvat (ed.) *Herencia, Medio y Educación*, Buenos Aires, 1974). In connection with this subject J. Ajuriaguerra says: 'It has been demonstrated that stimuli are essential for the maturing of the neuronal systems. When external stimuli do not exist – or are insufficient – the organization of the activity of the cerebral cortex is retarded or functions incorrectly.' (Ajuriaguerra, J., *Manual de Psiquiatría Infantil*, Barcelona, 1973)
13 See, for example, the following figures taken from Garret, H., *Great Experiments in Psychology*, New York, 1951.

Fig. 2.3 *Changes in mental capacity with age*

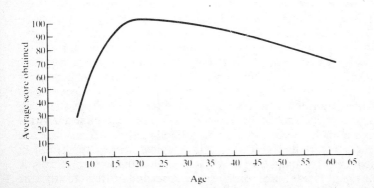

Fig. 2.4 *Reduction of score in the intelligence test after puberty*

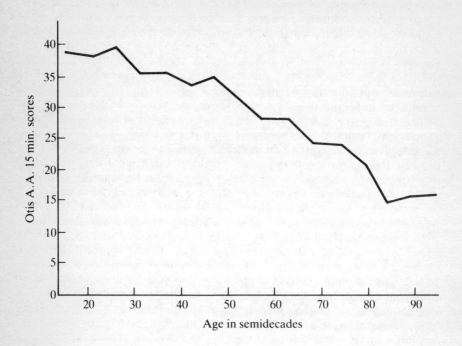

14 Nevertheless, the manifestations of a real integral maturity growth also appear at the corporal level (for example, in greater plasticity and liberty of movements, enriching of expressive possibilities through gestures, facial expressions, etc.
15 For this concept see Rapaille, G., *Laing*, Paris, 1972.
16 For this concept see Janov, A., *The Primal Scream*, New York: The Putnam Publishing Group, 1970, particularly chaps. 2, 3 and 4.
17 In connection with this, see Goffman, E., *The Presentation of Self in Everyday Life*, New York: Doubleday & Co Inc., 1959
18 On the concept of authoritarian personality see the classical work by Adorno, T. W., et al., *The Authoritarian Personality*, New York: W. W. Norton & Company Inc., 1950.
19 On the concept of rational and irrational authority, see Fromm, E., *Man for Himself*, New York: Ballantine Books, 1965, chap. 2, paragraph 1.
20 Fromm, E., *The Sane Society*, New York: Ballantine Books, 1983, chap. 5.
21 It would be interesting to study the structural conditions (economic, political, and institutional) which make possible or perturb the existence of what are here described as 'healthy groups'. The structural characteristics of a society mould the groups directly or indirectly in some way. A markedly authoritarian political structure would have the effect of reproducing inside the small groups the domination relationships through the distribution of hierarchical roles. If it should be possible to constitute a healthy group within a culture which does not encourage it naturally, the

negative influence of the medium will bombard it continually, sometimes subtly, sometimes more openly.
22 For this reason, the sociologist approach which reduces the problem of psychological conditioning to a series of economic-social constraints is insufficient to explain the different ways that the psychological history of a real, concrete individual may take.
23 For a list of description of these features, we refer the reader to the chapter 'Small Groups and Personal Growth: Distorting Mechanisms versus the Healthy Group' in this volume.
24 See, for example, one of the consequences of an inadequate environment: 'When protection against the cold is not obtained by normal means (clothing and comfortable dwelling) the organism resorts to an increase in the production of internal heat by burning up substances received with food or those held in reserve in certain corporal structures. Since the inadequate conditions of housing and clothing of large sectors of the population are generally accompanied by deficient energy-producing diet, the organism loses its capacity for protection against the cold and only the reduction of muscular activity and prolonged periods of repose enables the energy equilibrium of the individual to be relatively maintained. The economic and social consequences originating from this situation, which constitutes a truly vicious circle, have a profound significance . . . In babies, the physiological needs of nutrients, especially proteins, increase as a result of the frequent bouts of infectious diseases which cause a reduction in ingestion and a greater metabolic deterioration. If this deficiency cannot be compensated with higher than normal ingestion after recovery from the disease, a progressive deficit occurs in each episode. Additionally, parasitic infections and diarrhoetic illnesses diminish the intestinal absorption of nutrients essential to the diet. The combined effect of these factors is a greater need for nutritional sources and/or calories, so that the rations of pre-school age children, especially those from the lowest social levels, ought to be considerably greater than those calculated for a normal situation.' (Sabulsky and Battellino, 'El Problema de la Alimentación Humana', *Ciencia Nueva*, 23, Buenos Aires, 1973).
25 As Fromm points out in *The Sane Society*, it cannot be claimed that the material progress of the developed nations has brought with it true progress in psychological balance and growth: 'We see, moreover, that the most democratic, pacific, and prosperous countries of Europe, and the most prosperous country in the world, the United States, present the most serious symptoms of mental disturbance. The objective of all the socio-economic development of the Western world is to have a materially comfortable life, a relatively equitable distribution of wealth and stable democracy and peace. Yet these same countries which have come closest to that objective show the gravest symptoms of mental unbalance. It is true that these figures in themselves prove nothing, but at all events they are surprising . . . Is it possible that the life of prosperity which the middle class leads, while satisfying our material needs, should leave us a sensation of profound boredom and that suicide and alcoholism should be the pathological ways of escaping from this boredom? Is it possible that these figures constitute a radical illustration of the truth of that saying "Man does not live by bread alone" and reveal that modern civilization does not satisfy some profound needs of the human individual?'

3
On Human Development, Life Stages and Needs Systems

Carlos A. Mallmann

Introduction

Our purpose in this chapter is to contribute to the long-term task of reaching a theoretical understanding of human development which takes into account simultaneously universal and particular human structures and processes, culturally and historically-determined social dimensions, and natural and human-made habital characteristics.

Our contribution focuses on some of the human and social aspects of this problematique. It is based on the relationship between two fields of research which have been quite unrelated in the past, namely, human life-stages theory and human needs theory.

Human life-stages

Psychologists, social psychologists, anthropologists and biologists have often described the development of human beings as a sequence of qualitatively different life stages. They maintain that while human development is basically continuous it also proceeds in distinct spurts, and that these tend to have a universal epigenetic ground plan.

Very well known are the physiological development stages or spurts. Good examples are the growth spurts during the gestation and adolescent stages, women's menarche and menopause and the appearance of deciduous, permanent and wisdom teeth. Others are less well known. The existence of a universal, human epigenetic physiological ground plan is generally accepted.

A universal human epigenetic psychosocial ground plan was proposed by E. H. Erikson.[1] In this scheme he integrates Freud's psychosexual development stages and says that one of the purposes of his work is to facilitate the comparison with other schedules of development (physiological, cognitive, etc.).

We base our work mainly on Erikson's results because they cover the whole human life-span, they are based on observations in different cultural settings, they show the relation between stages and basic elements of society and they introduce the continuum which goes from human mal-development to human well-development.

More recently H. B. Green has proposed that the human being self-develops by solving a succession of eleven temporal problems encountered in its life-span.[2] The solution of each one forms a new state in the development of the self-in-time. The life-span of the self-in-time can be divided into stages, each one normally occurring within an age-range when the new self first begins to change through achieving a new time orientation.

Other authors have looked more at the stages of childhood and adolescence: Piaget at cognitive development,[3] Kohlberg at moral development[4] and Gesell at motor development.[5]

We have recently been shown evidence that human psychophysiological stages tend to be equally spaced in a sensorial-age scale.[6] We propose that sensorial-age is proportional to the square root of the chronological-age of a person measured from conception onwards and that the proportionality constant is the inverse of the square root of the chronological gestation period (0, 7 years). As a consequence, sensorial-age is given by a pure number which is zero at conception, one at birth, five at 16.8 years of age and ten at 69.3 years of age.

In Table 3.1 we show the comparison of Erikson's psychosocial stages and Green's temporal stages with some well-known physiological development stages and C. Show Schuster and S. Smith Achburn's phases of life cycle in a graph where life-lapses are represented linearly in units of sensorial-age.[7] The equivalent chronological-ages and their errors are indicated as well.

The Table shows that the growth spurt of the foetus takes place in the first stage, the growth spurt of adolescence, women's menarche and men's testis development take place in the fifth stage, that women's menopause takes place in the ninth stage, that deciduous and permanent dentition take place in the second and fourth stages respectively and that wisdom teeth appear in the sixth stage. There is an excellent correlation between these physiological changes and the equally-spaced sensorial-age life-stages.

The psychosocial development stages and the psychotemporal stages, both of them more loosely defined timewise than the physiological ones, also tend to correlate with the assumed life-stages, as can be seen in the Table. An exception is the autonomy-versus-shame psychosocial stage of Erikson which covers part of the second and third stages. The stages after the one of identity-versus-role

Table 3.1 Human Life-Stages

Polar Potentialities Actualization Stages	Green's Psycho-Temporal Stages	Erikson's Psycho-Social Stages
Continuation to Decline		
Peace to Despair	The Rich Past	Ego Integrity versus Despair
Synergy to Antagony	Foreshortened Future	
Variety to Monotony	Reconsidered Time	
Action to Inaction	The Uses of Time	Generativity versus Stagnation
Relating to Isolating	Alternatives in time / Mutual Time	Intimacy versus Isolation
Meaning to Lack of Meaning	Personal Time	Identity versus Role Diffusion
Understanding to Lack of Understanding	Causal Sequences	Industry vs. Inferiority
Autonomy to Dependency	Restricted Time / Clock Time	Initiative versus Guilt
Security to Insecurity	Permanence of Objects	Autonomy vs. Shame / Trust vs. Mistrust
Existence to Inexistence		

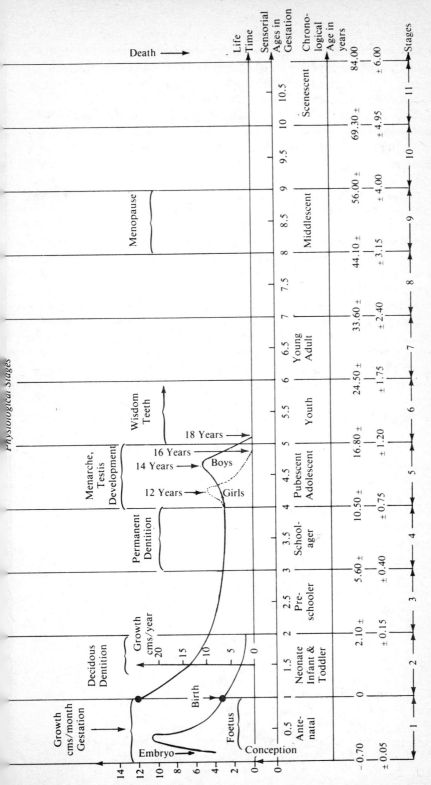

diffusion cannot be located well because their age spans were not given by Erikson. In the case of Green's temporal problems, the exceptions are clock-time which covers part of the second and third stage, mutual-time which covers part of the fifth and all the sixth stage and uses-of-time which covers the seventh and eighth stages.

The above analysis leads us to submit the working hypothesis that the psychophysiological development stages start at conception and are given by equally-spaced sensorial-age life-spans whose duration is of one sensorial gestation.

In the Table we also introduce polar potentialities actualization stages which we shall describe and analyze in the next section. They allow a relation to be established between life-stages theory and needs theory.

According to Erikson the universal epigenetic human development ground plan unfolds as the person relates with himself and with his successive social and 'habital' environments, giving it its personal social and habital specificity. (We introduce the word 'habital' by which we mean: of or relating to the human habitat, paralleling the meaning of the words social and personal. We are not using the word ecological because in its usual meaning it does not include the human-made habitat.) In this respect, it is his view that 'at each stage one more nuclear conflict adds a new ego equality, a new criterion of accruing human strength' and that 'each comes to its ascendence, meets its crisis, and finds its lasting solution during the stage indicated' but never forgetting that 'all nuclear crises must exist from the beginning in some form, for every act calls for an integration of all'. The eight universal nuclear conflicts he proposes are, in the order they appear during life, the ones given in Table 3.1. 'An individual, or a culture, may linger excessively or move in an accelerated progression over one or more of them. This makes room for variations in tempo and intensity. Each such acceleration or retardation, however, is assumed to have a modifying influence on all later stages. An epigenetic diagram thus lists a system of stages dependent on each other . . .'

When the lasting solution of each nuclear conflict results from a 'favourable ratio' of its positive and negative elements, according to Erikson, the human outcomes are the strengths and basic virtues (in italics) mentioned below after the polar extremes of the nuclear conflicts:

1 Basic trust versus basic mistrust: drive and *hope*;
2 Autonomy versus shame: self-control and *willpower*;
3 Initiative versus guilt: direction and *purpose*;
4 Industry versus inferiority: method and *competence*;
5 Identity versus role confusion: devotion and *fidelity*;
6 Intimacy versus isolation: affiliation and *love*;
7 Generativity versus stagnation: production and *care*; and
8 Ego integrity versus despair: renunciation and *wisdom*.

These outcomes are what we could call human well-development. Obviously when the lasting solution of each nuclear conflict results from an unfavourable ratio of its positive and negative elements we shall be in the presence of human mal-development.

Another essential feature of Erikson's scheme is his proposed 'relation of each successive stage and crisis to one of the basic elements of society and this for the simple reason that the human life-stages and man's institutions have evolved together . . . society, in principle, tends to be so constituted as to meet and invite this succession of potentialities for interaction and attempts to safeguard and to encourage the proper rate and the proper sequence of their unfolding.'

Some of the results of the human life-stages theory having been described very succinctly, two questions arise immediately: What is the relation between it and the human needs theory? And is the epigenetic human development ground plan universal? These will be the topics of the next two sections.

System of human needs

In order to be able to relate Erikson's nuclear crisis with human needs, we have to introduce the concept of human polar potentialities. It is well known that human beings have potentialities which go from one extreme of a continuum to the other. Examples of polar potentialities are those which go from existence to inexistence, security to insecurity, autonomy to dependency, understanding to lack of understanding, relating to isolating, action to inaction, etc. The requirements which human beings have for actualizing the extremes of those polar potentialities which lead to their state of health are called human needs.[8] Each polar potentiality has a corresponding need.

Examples corresponding to the potentialities given above are the needs for existence, security, autonomy, understanding, relating and action. Since all human needs are interrelated, they and also polar potentialities form a system.

Polar potentialities are actualized by human beings through their successive actions in life and become evident in the form of their character structure; desires, aspirations or wants; and drives, motivations or conations. They are the human beings' particular actualities corresponding to their universal polar potentialities. There are human beings who do not have all the potentialities due to some irreversible cause and they are called disabled. When the cause is reversible, attempts are made to correct it.

A different situation is that of human beings who have all the potentialities but actualize one or more of the extremes of the polar potentialities which lead to a state of illness. In this case, efforts are made to identify the combined social, habital and personal causes which lead them into such a situation, and to take the measures which reverse them.

It is our contention that Erikson's nuclear conflicts are the way in which he expressed the fact that human potentialities have this polar structure. His 'favorable ratio' of the nuclear conflicts' positive and negative elements in the solution of them is expressed in needs theory by saying that the parts of the polar potentiality continuum which will manifest themselves in a person's character, desires and drives depend on the way, positive or negative, in which potentialities have been and are being actualized. If this is so, we now require to correlate his nuclear crisis with a list of polar potentialities or, in other words, a list of human needs.

We shall use our system of human needs as a guide and adapt it for this purpose.[9] We suggest the correlation of nuclear crises and potentiality categories which are obtained if the following potentiality categories are assigned to the stages of Table 3.1.

Stage Categories of polar potentialities continua which go from:

1	Existence to Inexistence
2	Security to Insecurity
3	Autonomy to Dependency
4	Understanding to Lack of Understanding
5	Meaning to Lack of Meaning
6	Relation to Isolation
7	Action to Inaction
8	Variety to Monotony
9	Synergy to Antagony
10	Peace to Despair
11	Continuation to Decline
12	Declination to Death

The first nine potentiality categories correspond to the following need categories of our previous classification: (1) Maintenance, (2) Protection, (3) Autonomous Participation, (4) Understanding, (5) Meaning, (6) Love, (7) Creation, (8) Recreation, (9) Synergy.

The adaptation forced us to introduce some changes which we think are positive. The category Autonomous Participation was transformed into Autonomy and the category Love into Relation. By so doing, we have separated clearly the need for autonomy from the need for participation which we have now included together with love as one of the sub-needs of the category Relation.

The other two categories whose names were changed are Creation and Recreation which became Action and Variety respectively. In both cases we now have a category which is more general than, and which includes, the previous one.

Another effect of the adaptation is a change in their order of presentation. Now the order has a meaning. The successive polar potentialities have priority over the other ones at the corresponding life-stages. In Table 3.2 we show the system of human needs which corresponds to the adapted potentiality categories together with a way of grouping them further and the ages and stages in which the needs categories have priority. Erikson's statement that all nuclear crises must exist from the beginning in some form, for every act calls for an integration of all, is expressed in needs theory by the statement that potentialities form a system in which all of them are interconnected and, as a consequence, present in every act.

Having sketched the answer to the question about needs theory, we can now look into the question about the universality of the epigenetic ground plan.

Epigenetic ground plans

According to the universal epigenetic human development ground plan theory, each life-stage has one nuclear crisis which has priority over the others. In needs theory we would translate this statement by saying there is a universal diachronic hierarchy of potentialities which unfolds during life. This means that at each stage's age-span all the potentialities are present but are mainly used through their inter-relations, to satisfy the corresponding priorized need category. This statement can be represented in a two-dimensional graph, in which the polar potentialities categories are plotted on the vertical axis and the stages or age-spans on the horizontal axis.(The use of this kind of graph was suggested to me by Johan Galtung.) Table 3.3 shows such a graph.

The Xs indicate the priorized polar potentiality category for each life-stage according to the universal epigenetic human development ground plan theory, and the Os the fact that all other polar potentialities categories are present at each stage. This kind of graph is the translation into needs theory of Erikson's epigenetic chart or matrix.

If the plan were not universal but instead culturally determined, any trajectory, not necessarily the diagonal one, would be possible in principle. More than that, one could then imagine very much more 'blurred' trajectories instead of a one-to-one relationship between life-stages and potentiality categories. All of this could be represented in the same kind of graph by assigning first, second, third, etc. priority to the different polar potentiality categories in each stage. Each culture would then have its characteristic graph. We believe that the available evidence is in favour of a universal ground plan but it does not completely exclude other possibilities, particularly a more blurred one around the diagonal. More anthropological research on this subject is required.

Even if the ground plan were universal, there exists a strong cultural

Table 3.2

Classification of needs according to their appearance in successive life-stages	Classification of needs according to categories of satisfactors which mainly satisfy them	Personal	Extra-personal	
		Psycho-somatic	Psycho-social	Psycho-habital
−0.70 ± 0.05	Conception			
0	Stages Birth			
	1 Existence	Nutrition Rest Exercise	Social Habitability	Physical Habitability
2.1 ± 0.15	2 Security	Prevention Cure Defence	Prevention Restitution Defence	Prevention Restitution Defence
5.6 ± 0.40	3 Autonomy	Self- Determination	Independence Liberty	Independent Habitat Use
10.5 ± 0.75	4 Understanding	Psycholization Introspection Study	Socialization Education Information	Habitatization Observation
	5 Meaning	Self- actualization Goals	Historic Prospective Religious Co-actualization goals	World View, View of Nature

Stage brackets (right side):
- 1: self-subsistance
- 2: self-assurance — Self-affirmation
- 3: self-reliance
- 4–5: self-orientation

ON HUMAN DEVELOPMENT, LIFE STAGES AND NEEDS SYSTEMS

Age	Stage	Phase	#	Need	Personal	Social	Habital
16.8 ± 1.20	co-living	Co-affirmation	6	Relating	Self-belief, Self-love	Friendship, Love, Participation	Rooting, Attachment
24.5 ± 1.75	co-operating		7	Action	One's own creations (intellectual, artistic, etc.)	Creation of Social Valuables	Creation of Habital Valuables
33.6 ± 2.40	co-experiencing		8	Variety	Self-recreation, Self-change	Social recreation, Social variety	Habital recreation, Habital variety
44.1 ± 3.15	co-developing		9	Synergy	Authenticity, Equanimity, Security, Humility	Solidarity, Justice, Altruism, Generosity, Responsibility	Beauty, Ecological Equilibrium
56.0 ± 4.00	maturing	Trans-affirmation	10	Peace	Personal	Social	Habital
69.3 ± 4.95	comprehending		11	Continuation	Wisdom	Wisdom	Wisdom
84.0 ± 6.00	transcending						

— Death

Table 3.3 *Epigenetic human development ground plan*

Categories of polar potentialities – Continua which go from:

	1	2	3	4	5	6	7	8	9	10	11
Continuation to Decline	o	o	o	o	o	o	o	o	o	o	X
Peace to Despair	o	o	o	o	o	o	o	o	o	X	o
Synergy to Antagony	o	o	o	o	o	o	o	o	X	o	o
Variety to Monotony	o	o	o	o	o	o	o	X	o	o	o
Action to Inaction	o	o	o	o	o	o	X	o	o	o	o
Relation to Isolation	o	o	o	o	o	X	o	o	o	o	o
Meaning to Lack of Meaning	o	o	o	o	X	o	o	o	o	o	o
Understanding to Lack of Understanding	o	o	o	X	o	o	o	o	o	o	o
Autonomy to Dependence	o	o	X	o	o	o	o	o	o	o	o
Security to Insecurity	o	X	o	o	o	o	o	o	o	o	o
Existence to Inexistence	X	o	o	o	o	o	o	o	o	o	o

Life Stages →

and personal determination of the successive nuclear crisis outcomes because of the different ways each society, community and family actualizes the human potentiality system and the different inherent characteristics of each person. As a consequence, human beings' values, priorities assigned to polar potentiality categories and within them to parts of its continuum, satisfactor preferences and hence, desires, aspirations and drives are not at all universal. The way in which a potentiality category is actualized in one society can be unacceptable in another. The same statement is true for persons within a society.

We now have to say a few words about the relation of this diachronic needs hierarchy and Maslow's synchronic needs hierarchy.[10] What he has stated is that if you deprive human beings of the level of satisfaction of needs they have previously achieved, they will reorient their activities in a way which is consistent with the following hierarchy of needs: physiological, safety, belongingness and love, esteem, self-actualization, understanding and knowledge. It answers a different kind of question from that which we are answering with the diachronic needs hierarchy. The question, then, is if our scheme has something to say about his question. We would say that the answer is open in our scheme and that one has to introduce additional assumptions to be able to answer it theoretically or use the existing empirical evidence about it.

Our scheme certainly introduces an additional dimension into Maslow's question, namely, the age-span or life-stage of the person who is being deprived. A first working hypothesis could be that the synchronic needs hierarchy at each age is the same as the diachronic one. It would agree with Maslow's hierarchy only for the first two needs. More research results are required to settle these questions.

Having made an exploration of the interrelation of the life-stages human development theory and needs theory, we can return to the main topic of our chapters, which is human development.

Human development

We can now show how the foregoing thoughts contribute to the development of the topic of human development which T. Barreiro elaborated in her contribution to this book.

We answer her question, Does authentic or ideal growth exist? by saying that Yes, it consists, as Erikson says, of solving the sequence of nuclear crises positively, which means with a favourable ratio between the two opposites of the crisis. In needs theory we would say: by actualizing the potentialities in such a way as to have the positive parts of the polar potentiality continuum, in other words, needs, manifest themselves as desires, aspirations and drives. These statements imply that we can, with the same theory, understand mal-development. It consists of actualizing the potentialities in such a way as to have the negative parts of the polar potentiality continue to manifest themselves.

Her classification into primary and maturity growth can be seen in our scheme as the division of our stages into two groups. The primary growth corresponds to our stages 1 to 5, and the maturity growth to our stages 6 to 10. The statement that maturity growth is rooted in primary growth can be seen as a consequence of Erikson's assertion that the solution of each nuclear crisis is affected by the way the previous ones were solved.

As to the tasks required to be undertaken in order to build a model of human development, we can say the following:
(a) by interrelating the fields of human life-stages, of human potentiality and needs systems and of human development, we can use the great wealth of empirical knowledge available in these fields to make explicit the model of human development prevailing *de facto* in different cultures, the model which feeds and is fed by social dynamics;
(b) by continuing the research in these interrelated fields, we shall continue contributing to the task of drawing up a model of authentic human development;
(c) the latter tasks will progressively allow the deepening of the

comparison between the prevailing, culturally-determined, human development models with the model of authentic human development;

(d) by using the results of human life-stages and potentialities and needs systems research on the conditions (economic, social, political, habital, cultural) which enhance human development, we would contribute to determining the conditions of feasibility of the human development model.

The construction of a coherent theory which brings into one picture all of this knowledge remains to be done. What we have accomplished in this preliminary exploration is more to raise questions than to give answers. The work ahead is an important intellectual challenge for the future.

Notes

1 Erikson, E. H., *Childhood and Society*, New York: W. W. Norton & Company Inc., 1950, 1963.
2 Green, H. B., 'Temporal Stages in the Development of Self' in Fraser, J. T. and Lawrence, N. (eds.) *The Study of Time II*, New York: Springer Verlag, 1975.
3 Piaget, J., *Construction of Reality in the Child*, New York: Basic Books, 1954.
4 Kohlberg L. and Turiel E., *Research in Moral Development*, New York: Holt, Rhinehart & Winston, 1971.
5 Gesell, A., *First Five Years of Life*, New York: Harper & Row, 1940.
6 Mallmann, C. A., 'Notes on Sensorial Times, Time Views and Development Problematics', Fundacion Bariloche, Argentina, 1980.
7 *The Process of Human Development, A Holistic Approach*, Boston: Little Brown & Co., 1980, Table 1-1. Antenatal from conception to birth; Neonate, 0 to 28 days; Infant, 0 to 1.25 years; Toddler, 1.25 to 2.5 years; Preschooler, 2.5 to 5 years; School-ager, 6 to 10 years; Pubescent, 10 to 12 years; Adolescent, 13 to 18 years; Youth and Young Adult, 18 to 35 years; Middlescent, 35 to 65 years; Senescent, 65 and more years.
8 Mallmann, C. A. and Marcus, S., 'Logical Clarifications in the Study of Needs' in Lederer, K. (ed.) *Human Needs*, Cambridge, Mass.: Oelgeschlager, Gunn & Hain, 1980.
9 Mallman, C. A., 'Society, Needs and Rights: A Systemic Approach in Human Needs' in *Human Needs*, op. cit.
10 Maslow, A., *Motivation and Personality*, New York: Harper & Row, 1970 and *The Farther Reaches of Human Nature*, New York: Viking Press, 1973.

4

On Turning Development Inside-Out or (better) On Not Turning Development Outside-In in the First Place

W. Lambert Gardiner

1 From outside-in to inside-out

The distinction between a micro and a macro level of analysis is a very useful tool in our study of development, as it has been in the study of so many other phenomena. Those who study development at the micro level focus on the individual as their system, those who study development at the macro level focus on the institution as their system, and those who aspire to integration can interlock the micro and macro contributions by recognizing that they are dealing with a hierarchy of systems within systems in which the individual is a subsystem of the institution.

However, both those levels of analysis within traditional science represent the point of view of an 'objective' observer looking at the systems from the outside-in. They differ only in their relative closeness to the hierarchy of system – the micro person is taking a close-up view of the individual whereas the macro person is taking a long-shot view of the institution in which the individual dwindles to a dot in a frequency distribution.

I will argue here that it is necessary to look at development from the inside-out as well as from the outside-in, that the familiar distinction between micro and macro levels of analysis must be supplemented by the less familiar distinction between outside-in and inside-out points of view.[1] Those two dichotomies yield four approaches to development as diagrammed in Fig. 4.1. Tentative names for each of those four approaches have been entered in the appropriate cells.[2]

Fig. 4.1 *Four approaches to development*

	OUTSIDE-IN	INSIDE-OUT
MACRO	capital commun } ism social	communalism
MICRO	behaviourism	humanism

The micro-macro distinction encourages a strategy of integration in which the micro people work up to the macro level and the macro people work down to the micro level. I will argue here that they may suffer the fate of Pat and Mick who started building a tunnel by working from both ends toward the middle and ended up with *two* tunnels.[3] I will argue, further, that I can see light only from one end of the tunnel – the one which goes from micro to macro. This may simply be the rationalization of a micro man but I will try to present a rationale for this argument. I will argue even further that this light can be seen only when looking one way – from the inside-out rather than from the outside-in.

That is, I am abandoning the good-natured democratic stance that each level of analysis and each point of view is equally valid and saying that one should start with the micro level of analysis and inside-out point of view (that is, from the bottom right cell of the 2 × 2 matrix in Fig. 4.1).

However, most discussion of development is from a macro level of analysis and an outside-in point of view – that is, from within the top left cell of the 2 × 2 matrix in Fig. 4.1. One typical textbook, *Problems of the developing nations* by L. P. Fickett, is organized under the 'four fundamental aspects of the development process' – the sociological, the economic, the military, and the political (that is, all of them are at the macro level of analysis and the outside-in point of view).

One reason why discussion is at a macro level is clear. Those responsible for public policy tend to consult experts on institutions (economists, political scientists, sociologists, anthropologists) since analysis at this macro level helps clarify the 'big picture' within which they must act. One reason why discussion is from an outside-in point of view follows. Insofar as people can be seen at all, from this bird's eye level, it is their behaviour rather than their experience which is observed.

This distinction between behaviour and experience is crucial in understanding the outside-in inside-out dichotomy. The system studied by psychology – the person – is unique among all the systems in the universe in that it can be observed from the inside as well as from the outside. That is, were I able to observe you just now as you are reading this, I could observe you from the outside – that is, your behaviour.

However, you from your exclusive ringside seat, can observe yourself from the inside – that is, your experience.

In a textbook on educational psychology which I published recently, I tried to turn human development inside out – that is, to shift from the current behaviouristic emphasis on behaviour to a humanistic emphasis on experience. My argument in that book is summarized in the next two sections, in which the outside-in view of behaviourism (section 2) is contrasted with the inside-out view of humanism (section 3). Having succumbed to the temptation to summarize my book in one sentence (I find that this is the best way to deal with temptation), I am tempted to summarize social reform by humanists in a similar succinct sentence, suitable for transcription to a bumper sticker. Once again, I gleefully succumb.

HUMANISTS TRY TO TURN SOCIAL DEVELOPMENT INSIDE OUT

The two projects seem parallel – the former shifting from the bottom left to the bottom right cell in Fig. 4.1 and the latter shifting from the top right to the top left cell in Fig. 4.1. However, it is difficult for people working at a macro level of analysis to shift directly from an outside-in to an inside-out point of view, since, as argued above, the traditional macro level of analysis requires an outside-in point of view.

The two projects are complementary rather than parallel. My project could be seen as complementary to those of Thelma Nudler and of Johan Galtung. Nudler contributes to a more enlightened discussion of development by recommending that it be extended from the macro to the micro level – that is, from the top left cell to the bottom left cell in Fig. 4.1. In this chapter, I am recommending a second shift from the outside-in to the inside-out point of view within the micro level of analysis – that is, from the bottom left cell to the bottom right cell. Galtung recommends a third shift from a structure-oriented to an actor-oriented perspective on society or, in the terms used here, from the micro to the macro level of analysis within an inside-out point of view – that is, from the bottom right cell to the top right cell.

Since the direct route from the top left to the top right cell is difficult, I recommend the Nudler–Gardiner–Galtung scenic route. Go down Nudler Pass from top left to bottom left, take the Gardiner Expressway from bottom left to bottom right, and then go up Galtung Pass from bottom right to top right.

Those brave souls who allow themselves to be nudged by Nudler from top to bottom left must learn to ignore the catcalls of 'reductionist' from the loftier macro regions. Those who are goaded by Gardiner from bottom left to right must not only ignore the cries of 'reductionist' from above but also those of 'subjectivist' from the left. There should be a similar epithet (to convey the opposite of 'reductionist') for those who are further goaded by Galtung from bottom to top right.[4]

Whatever the epithets, the courage in countering them may be rewarded by a fresh view of social development, which is not available to those who are locked in that traditional macro outside-in top left cell. In section 2, The outside-in view of behaviourism, I shall present a view of development from within the bottom left cell, in section 3, The inside-out view of humanism, I shall move into a view of development from within the bottom right cell, and in section 4, From micro to macro, I shall move into a view of development from within the top right cell. Though this second shift takes me outside the range of my professional competence, I shall try to share my glimpse of a light at the end of the tunnel. Those of us whose competence is at this macro level of analysis may then be encouraged to continue digging in this direction.

2 The outside-in view of behaviourism

(a) The person has only extrinsic needs

One broad question in psychology is 'What is the function of the nervous system?' and one broad answer provided by the theory of evolution is 'to enable the organism to survive?' The next question is '*How* does the nervous system enable the organism to survive?' and the classic answer is 'It ensures that the organism will approach things that are good for it (e.g. things that it eats) and avoid things that are bad for it (e.g. things that eat it).' The need-reduction theory describes the former mechanism and the activation theory describes the latter mechanism. The need-reduction theory and the activation theory are the means employed by behaviourists to fit psychology within the theory of evolution. Let us look briefly at each theory in turn.

You are alive. You are in a precarious state. Life is a narrow tightrope with death on either side. To stay alive, you must maintain yourself within a narrow range of temperature, blood-sugar concentration, metabolic rate, and so on. The process by which you maintain yourself in this optimal state, in which those indices are neither too low nor too high, is called homeostasis. Let us imagine you have just been wined and dined. You are in your optimal state. However, you cannot remain thus for long. The mere passage of time conspires against your bliss. You get thirsty. You get hungry. This physiological state of deprivation is called a need. The need can be satisfied by appropriate behaviour with respect to some appropriate thing in the environment (the goal) – drinking liquid in the case of thirst and eating food in the case of hunger. Since the nervous system is the only system which 'knows' the environment, the physiological need in the digestive system must be transformed into some psychological counterpart in the nervous system. This psychological counterpart (the drive) orients you to the appropriate goal. By making the appropriate response with respect to the goal, the drive is removed, the need is satisfied, and the optimal state is regained.

Let us now turn from the positive to the negative drives, from the tendency to approach things which are good for us to the tendency to avoid things which are bad for us, from the need-reduction theory to the activation theory.

There are two basic strategies for avoiding things that are bad for you. You can remove it or you can remove yourself. The first involves fight and the second involves flight; the emotion underlying the first is rage and the emotion underlying the second is fear. Consider one of our remote ancestors confronted by a sabre-toothed tiger. He can remove it or he can remove himself. He can kill it or run away. The only good tiger is a dead tiger or a distant tiger.

The need-reduction theory and the activation theory are diagrammed together in Fig. 4.2 to clarify the similarities and differences between them. Both theories involve a negative feed-back loop to maintain you in your optimal state. Both theories describe the nervous system as a mediator between the internal environment (that is, the other subsystems of the person) and the external environment. According to the need-reduction theory, the function of the nervous system is to mediate between a state of deprivation in the internal environment (a need) and a thing in the external environment that will satisfy the need (positive goal), so that you will approach that thing. According to the activation theory, the function of the nervous system is to mediate between a thing in the external environment (negative goal) and a state of the internal environment (an emotion), so that you will avoid that thing.

Since the nervous system is merely a mediator between internal and external environments, the person is extrinsically motivated.[5] The person is pushed and pulled by outside forces – pushed by needs and pulled by satisfiers of those needs, pushed by threatening things and pulled by the emotions generated by those things.

Behaviourists argue that all human motivation is indeed determined by those extrinsic needs. When asked to explain more sophisticated behaviours beyond those obviously linked to the primitive motivations of hunger and thirst and the primitive emotions of rage and fear, they talk of secondary drives, which are established through association with those primary drives. Thus, in the famous experiment in which chimps were taught to work for tokens which they could 'spend' in a chimpomat, their capitalistic tendencies were explained in terms of the tokens being the means of obtaining food to remove the hunger drive to satisfy the hunger need to regain the optimal state to survive. Capitalism is established by making money the means to the ends of satisfying those extrinsic needs.

Fig. 4.2 *Need-reduction theory (left) and activation theory (right)*

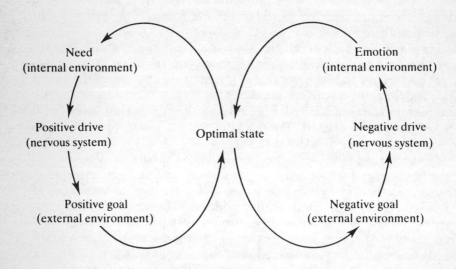

(b) The person is conditioned from the outside in

If the person has only extrinsic needs, then the person is conditioned from the outside in. There are two types of conditioning – classical and instrumental. Classical conditioning was demonstrated by Ivan Pavlov and elaborated by J. B. Watson, who tried to explain all behaviour within this paradigm. Instrumental conditioning was demonstrated by Edward Thorndike and elaborated by B. F. Skinner, who tried to explain all behaviour within *this* paradigm. Since those conditioning paradigms are very familiar (in contrast to the extrinsic motivation paradigm described above on which they are based), they are simply summarized here, side by side, in Fig. 4.3.

(c) The person is not responsible for behaviour

If the person is conditioned from the outside in, then the person is not responsible for behaviour. A person totally at the mercy of the environment is not free to act and thus not responsible for actions. Your behaviour today is determined by your conditioning in all your yesterdays. You are a victim of your past. You are depraved because you were

Fig. 4.3 *Conditioning, classical (a) and instrumental (b)*

(a) Classical conditioning - Pavlov and his dog.

The conditioned reflex is established by presenting the conditioned stimulus together with the unconditional stimulus which already elicits the response.

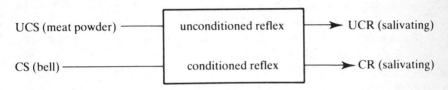

(b) Instrumental conditioning - Thorndike and his cat.

The conditioned reflex is established by arranging that the CR permits access to the UCS.

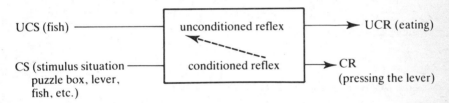

deprived. The compulsive consumer who buys a new car every year is not responsible, then, for the ultimate impact of this consumption on our planet. Nor is the person who sells them, nor is the person who produces them, nor is the person who drives the bulldozer that gouges the required metals out of the earth. None of them is any more responsible than the bulldozer itself. They are all as extrinsically motivated as it is.

(d) The person has only extrinsic worth

If the person is not responsible for behaviour, then the person has only extrinsic worth. He or she cannot be blamed for bad behaviour but, by the same token, cannot take credit for good behaviour. The person cannot then gain intrinsic worth and must thus seek extrinsic worth – that is, acquire possessions. The fact that such extrinsic worth is not an adequate substitute for intrinsic worth does not make it any less potent as a motivator. Indeed, it makes it *more* potent. Intrinsic worth is real and can thus be attained; extrinsic worth is an illusion and thus cannot be attained. It is therefore pursued more and more ardently. The person accumulates more and more possessions in a futile search for satisfaction. Thus consumption becomes compulsive.

I shall argue later that self-esteem is a genuine need of our species. This need will not go away simply because behaviourists do not recognize 'self' and hence 'self-esteem'. The person cannot gain self-esteem through possessions. You may be able to fool all of the people some of the time and some of the people all of the time, but you can't really fool yourself at all. The person attempts then to gain prestige (worth in the eyes of other people) rather than self-esteem (worth in one's own eyes). Prestige is not a satisfactory substitute for self-esteem. Other people are never really impressed by your possessions and, indeed, may be offended by them. They too may be seeking extrinsic worth by accumulating possessions. The bigger your pile, the less impressive their pile. Your greed clashes with their envy. Accumulation continues to escalate, however, because of – rather than in spite of – the fact that it is futile. Once you set out to impress, the less people are impressed, the more you try to impress them. The function of a thing is not its ostensible function, but to impress and thereby gain prestige. As many people as possible must know that you possess it. Thus consumption becomes not only compulsive but conspicuous.

(e) The person is an interchangeable part

If the person has only extrinsic worth, then the person is an interchangeable part. A person can, for instance, be fitted into a job. A job is defined by its functions, and the person who performs those functions can be replaced by any other person who performs those same functions.[6] Personnel officers fit the round pegs of people into the square holes of jobs. The person becomes defined by the job. People answer 'What do you do?' in terms of occupation, 'Who are you with?' in terms of organization, and 'What are you worth?' in terms of renumeration.[7] People with trivial jobs (that is, most jobs in an industrial society) see themselves as trivial people.

This further contributes to a lack of intrinsic worth and a compensating search for extrinsic worth. The fitting of the person to the Procrustean bed of the job also contributes to the irresponsibility mentioned earlier. If the driver of the bulldozer mentioned above ever had a twinge of conscience about his contribution to defacing the environment, he could rationalize it by saying 'It's my job.' Stanley Milgram discovered that people who would not normally hurt another person will do so when commanded by an authority figure. He called this the Eichmann effect, because of Eichmann's defence during his trial for exterminating Jews in a Nazi concentration camp, that he was merely doing his job as requested by his superiors. The crowd is less responsible than the individuals of whom it is composed. A corporation is a crowd.

(f) The person has contractual relationships

If the person is an interchangeable part, then the person has contractual relationships. The essence of a contractual relationship is that the people who make the contract are interchangeable. You have a contractual relationship with your grocer. There is an unwritten understanding that the grocer will give you food and that you will give the grocer money. It does not really matter to you that this particular grocer gives you the food and to the grocer that this particular customer provides the money.

You take the food home and cook it for your mate. It would seem reasonable to believe that your relationship to your mate is qualitatively different from your relationship to your grocer. The people involved are not interchangeable. It is important to Martha Jones that it is John Smith, the unique man she has chosen to live with, who is making her dinner, and it is important to John Smith that it is Martha Jones, the unique woman he has chosen to live with, who is sharing his bed. That is, some of our relationships are intimate rather than contractual. Intimate relationships are not possible, however, between people who are totally motivated by extrinsic needs. If there is no self, then what is there to be intimate with? The behaviouristic concept of the person implies then that your relationship to your mate is only quantitatively different from your relationship with your grocer. You simply present your mate with a longer and more complex shopping list consisting of psychic as well as physical goods. Your mate retaliates with an equivalent list. 'I'll scratch your back if you'll scratch mine.'

This cynical view of human relationships is not some 1984ish vision of a dehumanized world but a necessary deduction from the behaviouristic concept of the person. B. F. Skinner makes this explicit in his book *Verbal Behavior*. There are two ways you can get things done: you can do it yourself or you can ask someone else to do it for you. The former involves nonverbal behaviour and the latter involves verbal behaviour. Verbal behaviour is that behaviour by which you get things done through the mediation of other people. Other people are means to your ends.

The relationships between the person and the society are an extension of this relationship between any two members of society. The social contract is a set of explicit rules to ensure that all those implicit contracts between individuals are honoured. Government is a protection racket. You pay your taxes and they pay police to protect you from exploitation by other people. Government is an insurance company. You pay your taxes and they support you when you are old or sick or unemployed.

If relationships are contractual – that is, based on relative and arbitrary rules made by people to prescribe their conduct – those relationships are very tenuous. You are constantly apprehensive that the rules will be broken. You assuage your anxiety by being excessively careful that you meet your commitments in your contracts. Salvadore Maddi thus explains our obsession with doing our duty or, as it has come to be trivialized, doing our job.

3 The inside-out view of humanism

(a) The person has intrinsic needs

Your nervous system is an element of you as a person, and you as a person are, in turn, an element of your society. The nervous system has a very special role within this hierarchy of systems within systems. It is the only system which can 'know' your environment. It must know your environment in order to perform three broad functions – to mediate between your internal environment and your external environment (its biological function), to interact appropriately with other people (its sociological function), and to understand your environment and yourself (its psychological function). Underlying each of those functions are certain organic needs – biological, sociological, and psychological, respectively – which are designed to ensure that your nervous system performs each of those functions.

Biological needs
The biological needs and the mechanisms by which you satisfy them were described in section 2 (a) above. No evidence for such needs was presented, since no evidence is necessary. The best evidence for a need is that failure to satisfy it results in damage to the organism. If an organism is deprived of drink and food, it dies. Death is the dramatic documentation of the biological needs of thirst and hunger.

The humanistic concept of the goals of human development does not replace the behaviouristic concept but rather subsumes it. Humanism does not replace behaviourism as Copernicus replaced Ptolemy but subsumes it as Einstein subsumed Newton. Humanism accepts behaviourism as far as it goes but points out that it does not go nearly far enough. There are biological needs built into the nervous system, indeed, but there are sociological and psychological needs built in too. Since deprivation of those sociological and psychological needs results in less dramatic damage than death, it is necessary to present evidence that they exist.

Sociological needs
Just as the biological needs are based on the survival of the individual, so the sociological needs are based on the survival of the species. Mother Nature loads Jack and Jill not only with hunger and thirst drives so that each of them will survive as an individual, but with sex drives so that they will get together in order for our species to survive.

Since we can survive without sex – unhappy organisms but still live ones – we tend to assume that the sex drive is less powerful than the hunger and thirst drives. However, Mother Nature is more concerned with the preservation of the species than of the individual (like most parents, she aspires to be a grandparent) and would thus provide a powerful drive as a means to this end.

The sex drive is designed to ensure that Jack and Jill will not only get together for that delightful experience created to bribe us to procreate our species but that they will stay together to care for the resultant offspring during that long period of infant dependency. This caring mechanism is built into the child during this period of dependency so that the child will, in turn, take care of its child. Our co-operation with other people is based on this caring mechanism established within the family. Many modern evolutionary theorists argue that it is co-operation rather than competition that has made our puny species the King of the Jungle.

Total deprivation of sociological needs, like total deprivation of biological needs, also results in death. The human infant is so dependent that it could not survive without the care of other people. It is not ethically possible to study the effect of total sociological deprivation on human infants. However, Harry Harlow has tested the effect on our close cousin, the Rhesus monkey. Such deprived infants become neurotic, spending most of their time huddled in a corner of their cage. René Spitz has studied the effect of partial deprivation of sociological needs on human infants. He found that many infants raised in foundling homes with minimum social contact die. Those who survive are physically, emotionally, and intellectually stunted. They die a little bit. They fall short of becoming fully human.

Psychological needs
The psychological needs, unlike the biological and sociological needs, are not primarily concerned with survival. Our species is nature's first deluxe model with trimmings beyond those necessary for mere survival. We have more needs than we *really* need. The psychological needs reflect organic potentiality rather than organic requirements. They enrich rather than simply maintain life, they ensure that we thrive rather than merely survive, they make us competent in our environment rather than just adapted to it. As far as I can see, so far no one seems to understand why such luxury needs would evolve. Perhaps they evolved out of survival needs. We needed to know our environment in order to survive in it, but subsequently, as the threat to survival decreased, we needed to know our environment simply in order to know our environment. Psychological needs were means to an end but became ends in themselves. They became functionally autonomous.

How would you like $40 a day for lying in a comfortable bed doing nothing, with visors over your eyes, pillows around your ears, and cuffs around your arms, so that your leisure will not be disturbed? (Actually, it was $20 a day but those were 1956 dollars.) It sounds like a good deal. However, the McGill University undergraduates who were invited into this paradise for students were soon clamouring to get out. Such sensory deprivation turned out to be a very disturbing experience. Their thought processes deteriorated, their emotional responses became childish, and they had terrifying hallucinations. It seems that, just as the body needs food, so the mind needs stimulation.

This need for stimulation persists even when you are asleep. The discovery that rapid eye movements accompany dreaming has made it possible to conduct objective studies of this subjective state. Nathaniel Kleitman awakened some subjects every time they started to dream during several successive nights. On subsequent nights, during which they were allowed to rest in peace, they dreamed significantly more than before. When you are deprived of eating, you subsequently eat more; when you are deprived of dreaming, you subsequently dream more. You have a need to eat; you have a need to dream.

The satisfier of the need to eat is food, the satisfier of the need for stimulation is novel stimuli. Just as you seek food when you are hungry, you seek novel stimuli when you have a need for stimulation. A number of studies have demonstrated that organisms explore and manipulate their environment in search of novel stimuli. Rats will often choose the long scenic route over the short dull route between the start and goal boxes of a maze. They spend more time around unfamiliar objects put in their cage than around familiar objects. Monkeys will work hard to unfasten latches to open windows to see what is happening outside. Indeed, they will work to see nothing. They enjoy the manipulation of the latches as an end in itself. The activity is its own reward.

This need for stimulation may perhaps be explained in evolutionary terms. As long as your environment continues as it is, you are in no danger. It is only novel stimuli that are potentially dangerous. Exploration and manipulation of your environment makes the unfamiliar familiar. If the novel stimulus is indeed dangerous, you can remove it or remove yourself; if it is not dangerous, the threat is removed. Besides removing danger or threat of danger, exploration and manipulation incidentally enable you to get to know your environment. One peculiar property of novel stimuli may help explain why we know more than we really need to know. Over time, a novel stimulus becomes less and less novel; that is, it ceases to be a satisfier of the need for stimulation. We must therefore continually search for new stimuli to satisfy this need. Perhaps, as our environment becomes less and less threatening, this incidental function of getting to know our environment becomes more and more important.

A group of psychologists arranged to have some observers infiltrate an organization whose members believed that the world would end at a particular time on a particular date. They were curious to find out what would happen when that time passed and the world remained. The psychologists found that those people who were only peripherally involved with the group ceased to believe, whereas those people who were strongly committed to the group (that is, those who had stated their beliefs in interviews with the press, sold their belongings, cancelled their life-insurance policies, and so on) continued to believe. These true believers argued that the destruction of the world had been postponed because of rain, that the apocalypse had been cancelled because of their vigilance, that there had been a mistake in the date, and so on and so on.

These findings suggested to Leon Festinger, the leader of the group of psychologists, the concept of cognitive dissonance. When two items of information do not fit together, there is a tendency for one of them to be changed. For instance, the two items of information 'I smoke' and 'smoking causes cancer' are dissonant. Festinger found that fewer smokers than non-smokers believed the latter statement. People who have those two items of information within their subjective maps tend to stop smoking or stop believing. Research on cognitive dissonance has led to a number of further findings which Grandmother would not have predicted. Not only do we own a car because we read ads for it but we read ads for it because we own it; not only do we say what we believe but we come to believe what we say; not only do we own things we like but we come to like things we own; not only do we know what we like but we come to like what we know. All those findings point to a need for consistency.

Whereas the need for stimulation provides the organic basis for knowing our environment, the need for consistency provides the organic basis for understanding our environment. Not only do we need to know, but we need to know what we need to know. What we know must be organized into a consistent body of knowledge; that is, we need not only to know but to understand. The need for stimulation and the need for consistency together provide an organic basis for psychological growth. Jean Piaget describes the process of psychological growth as a series of alternating assimilations and accommodations. You assimilate information from your environment and adjust your subjective map of the environment to accommodate that information if it does not fit. The need for stimulation is the organic basis of assimilation and the need for consistency is the organic basis of accommodation. The need for stimulation ensures a fresh supply of new information and the need of consistency ensures that this information will be integrated within a consistent subjective map.

(b) Hierarchy of needs

Although the biological, sociological, and psychological needs must all be satisfied by the same nervous system, they are naturally in harmony. According to Abraham Maslow, they are organized in a hierarchy.

Biological needs are most potent; when they are satisfied, sociological needs become most potent; and, when those needs are satisfied in turn, psychological needs become most potent.[8] That is, you shift gears up the hierarchy as lower needs are satisfied. The biological and sociological needs are so easily satiated that, in a healthy person in a healthy society, most time is available for the less satiable psychological needs. Eating ruins your appetite. The satisfiers of sociological needs – namely, other people – are in plentiful (indeed, too plentiful) supply. The satisfaction of the survival needs provides pleasant periodic interludes from the rigours of satisfying the psychological needs.

The priorities within the hierarchy of needs make sense. The necessity of surviving (biological and sociological needs) comes before the luxury of thriving (psychological needs). Perhaps psychological needs could be seen in terms of surplus energy, much as our economic luxuries can be seen in terms of surplus capital. Individual survival (biological needs) comes before species survival (sociological needs) since it is the individual who is arranging the priorities. Mother Nature may be more concerned with the survival of the species, but I am more immediately concerned with my survival and you with yours. However, this hierarchy is not rigid. Sociological needs can take precedence over biological needs (as attested by the suicide and the martyr) and psychological needs can take precedence over sociological needs (as attested by the hermit and the monk).

I have expounded at perhaps too great length on the empirical evidence that there is an organic basis not only to biological needs, as argued by behaviourists, but also to sociological and psychological needs, as argued by humanists. However, such an exposition may be justified on the grounds that it establishes the empirical basis for replacing the basic proposition of behaviourism that 'The person has only extrinsic needs' with the basic proposition of humanism that 'The person has intrinsic needs'. It is this shift from extrinsic to intrinsic needs which is the foundation of the shift recommended in this chapter from the traditional outside-in point of view to an alternative inside-out point of view. The goal of human development, then, is not the satisfaction of biological needs in order to survive, as argued by the behaviourists, but the satisfaction of biological, sociological, and psychological needs to realize the full human potential built into the person at the moment of conception.

(c) The person grows from the inside out

If the person has intrinsic needs, then the person grows from the inside out. Every normal child has the potential to be fully a person, just as every normal acorn has the potential to be fully an oak tree and every normal kitten has the potential to be fully a cat. Powered by the intrinsic system of needs described above, the child seeks satisfaction for them. In an appropriate environment, children are able to satisfy those needs and thus realize their human potential. The basic project of the child is to become an adult – not any old adult but a great and good adult. We therefore need to 'explain' not the genius of a Pablo Picasso or a Margaret Mead (or whoever you think has most fully realized the human potential) but rather why we are not *all* Picassos or Meads. We need to explain not growth itself (this is simply the unfolding of the intrinsic potential) but the stunting of growth. Here are two alternative explanations.

Conflict between needs – Freud
The process of realizing the human potential is so long and so complex that many things can go wrong. It is relatively easy for an acorn to become fully an oak tree and for a kitten to become fully a cat, but it is not so easy for a child to become fully a person. The theory of Sigmund Freud could be considered as a dramatic documentation of the many things which can go wrong.

His id, superego, and ego represent the forces striving for the satisfaction of biological, sociological, and psychological needs, respectively. Although those forces are naturally in harmony, as argued above, Freud shows how they come into conflict.

Since the ego tries to maximize truth and since the id tries to maximize pleasure, they come into conflict when truth and pleasure are incompatible. Berelson and Steiner, in summarizing scientific findings about human behaviour, describe the human being as 'a creature who adapts reality to his own ends, who transforms reality into a congenial form, who makes his own reality'. In the conflict between truth and pleasure, it seems then that pleasure usually wins.

Since the ego is concerned with laws and the superego is concerned with rules, they come into conflict when laws (propositions created by humans to describe the world) and rules (propositions created by humans to prescribe their conduct in that world) are incompatible. Studies of conformity suggest that, in the conflict between laws and rules, rules usually win.

The gospel according to Freud is, therefore, that the attempt by our ego to know and understand the world is continually sabotaged by the id which chants 'I want' and by the superego which preaches 'Thou shalt not'. Any accuracy in our subjective maps of the objective world is a limited, hard-earned, and precarious accomplishment. This will remain the case unless we can design a world in which truth is invariably pleasant and rules are invariably rational.

Lack of satisfiers of needs – Maslow
Abraham Maslow argues that so many of us fail to realize our full human potential, not because our needs are necessarily in conflict or because conflict is artificially introduced, but because we fail to shift

gears up the hierarchy of needs. Most of the people on our planet must spend most of their time seeking satisfaction of their survival needs and have little 'spare time' for the luxury of seeking satisfaction for their psychological needs.

In our affluent industrial society, most of us have little direct experience of a subjective map in which biological needs are prepotent. We get an occasional glimpse of such a state when we are hungry and notice that we are highly sensitized to stimuli related to food. A psychologist once flashed nonsense syllables on a screen before lunch during a convention and got significantly more food-related responses than when he performed the same experiment *after* lunch. Volunteers in an experiment on the effect of semistarvation reported that their consciousness became dominated by food. They talked about food, dreamed about food, replaced the pin-ups in their lockers with photographs of food, and exchanged recipes rather than jokes with the other volunteers. Audrey Richards, an anthropologist, reports than food dominates not only the conscious life but the unconscious life of the members of an African tribe for whom food is very scarce. We are preoccupied not by sex, as Freud argued, but by whatever happens to be scarce – for example, sex in nineteenth-century Victorian Austria. The famished man does indeed live by bread alone.

Many of us in our affluent industrial society have more direct experience of a subjective map in which sociological needs are prepotent. Few of us get stuck at the level of biological needs, but many of us get stuck at the level of sociological needs. Other people – the satisfiers of sociological needs – are, as I said above, in plentiful supply. Perhaps so many of us get stuck in this second gear of sociological needs because of some distortions in social relationships within industrial societies – the tendency to seek prestige rather than self-esteem (see section 2(c) above), to establish contractual rather than intimate relationships (see section 2(e) above), and so on.

Maslow suggests that any benefit gained from visiting a therapist may be due to the fact that he or she satisfies sociological needs and thus permits the client to move up the hierarchy of needs. In our impersonal society, we need professional listeners to perform a function which is served by intimates in a traditional society. Many of us fail to satisfy our sociological needs and thus to move up the hierarchy to psychological needs and thus, thereby, to realize the full human potential.[9]

Maslow suggests further that the need to know may be overcome by the fear of knowing. The person is torn between the safety of the survival needs (biological and sociological) and the growth of the psychological needs. This conflict can be vividly illustrated by the image of a child clinging to a mother's apron strings in a strange environment, venturing out for longer and longer forays further and further into that environment, and dashing back to base after each foray when the environment gets too threatening. This conflict between the contentment of safety and the excitement of growth continues, in less blatant forms, throughout our entire lives. Many of us fail to venture far from our mother's skirts (or whatever symbolic equivalent we have established – tenured position, corner bar, suburban castle, etc.), whereas some of us do venture far – Neil Armstrong got all the way to the moon without his mother.

This emphasis on deficiency motivation at the expense of growth motivation – or, in Jerome Bruner's terms, on defending behaviour at the expense of coping behaviour – often results from a threatening environment. Dramatic changes in a person's competence when shifted to a non-threatening environment may be attributable to a switch from deficiency to growth motivation or, alternatively, from defending to coping behaviour.

(d) *The person is responsible for behaviour*

If the person grows from the inside out, then the person is responsible for behaviour. Extrinsic motivation requires extrinsic control, whereas intrinsic motivation requires intrinsic control. The constraining force of society on a person with extrinsic motivation may be replaced by the restraining force of a person with intrinsic motivation.

This is reflected, within the discipline of psychology, by a shift in emphasis from other-control to self-control. The lay person has been appropriately apprehensive about the psychologist because of the threat that, the better the psychologist understands you, the easier the psychologist can control you. This public image of the psychologist is somewhat justified, since the emphasis has indeed been on how to make organisms – whether rats or racoons, pigeons or people – behave as the behaviourist wants them to behave. The recent burgeoning of research on self-control (once a taboo topic) is an encouraging sign that psychology is turning from yet one more instrument of oppression to one of liberation.

This does not mean that we must start again from scratch. Each technique for understanding and control of other people is also a technique for self-understanding and self-control. Psychologists are beginning to present the powerful tools they have developed to the public so that people may use them for self-understanding and self-control. Even behaviouristic means can be used to humanistic ends. Power to the person.

This discussion of other-control versus self-control is a modern rephrasing of the determinism versus free-will chestnut which philosophers have been roasting for centuries. Perhaps this controversy has been with us so long because both positions are true. The determinist is determined because he believes he is determined whereas the free-willist has free will because he believes he has free will. The first act of free will is to believe in it.[10]

Two propositions which are incompatible with respect to the objective world may be perfectly compatible with respect to two different subjective maps of the objective world. People who have a behaviouristic self-concept based on extrinsic motivation will tend to believe that their behaviour is determined. The self-fulfilling prophecy (what you expect is what you get) will ensure that their behaviour is indeed determined. People who have a humanistic self-concept based on intrinsic motivation will tend to believe that they control their own behaviour. The self-fulfilling prophecy will, in this case, ensure that they indeed have free will. In this way, the determinist and the free-willist have both accumulated 'evidence' for their respective theories throughout our history. Each theory tends to be based on what feels good rather than on what seems true.[11] You do not believe it because it is true, but it becomes true because you believe it.

(e) The person has intrinsic worth

If the person is responsible for behaviour, then the person has intrinsic worth. We must accept blame for our bad behaviour but can take credit for our good behaviour. The words 'bad' and 'good' tend to scare scientists into scurrying off in search of philosophers. There seems no place for values in a world of facts. Western philosophers offer us a choice between pragmatic values (doing well) and ethical values (doing good). Some thinkers have, however, been evolving an alternative set of values based on natural laws rather than on cultural rules – that is, based on the propositions we have devised to describe ourselves and our planet rather than on propositions we have devised to prescribe our conduct on the planet. Here is a summary of that system of values, as expounded by such diverse thinkers as Teilhard de Chardin, Kenneth Boulding, and R. Buckminster Fuller.

Our species on our planet is confronted not so much with an energy crisis as with an entropy crisis. Since energy can neither be created nor destroyed, we have as much energy today as we ever had or ever will have. It is entropy – the natural tendency of a system toward disorder – that is increasing. Any process that destroys structure or breaks complex systems down into simpler systems contributes toward this spontaneous tendency of the universe toward chaos. Biological systems, within their limited space and for a limited time, defy this law of entropy. During the period of growth, they become more rather than less structured. Our species, the most complex biological system, is the greatest antientropic force in the universe. Each of us is a defiant little package of antientropy fighting our brave battle against the forces of chaos. Consciousness emerges as a function of complexity and serves as the ultimate weapon against entropy. It enables us to assimilate and accommodate information to create a microcosm of the universe within us. The more accurate this subjective map of the objective world, the better we fight the good fight.

It is a futile battle, because eventually we must submit to the forces of chaos. However, though it is futile for each of us as individuals, it is not futile for all of us as a species. Each of us spawns other little packets of antientropy in our books or our movies or our children or in the memories of our friends, before we are recycled in the air our survivors breathe and in the water they drink.

People who have this system of values recognize that they are a part of nature and not apart from nature. Since they are an important element in the complex system of the universe (and since the continuing functioning of the universe according to its natural laws is presumed to be a Good Thing), they have intrinsic worth. Their criteria of success is not wealth but health. They are healthy insofar as they realize their function in the universe – to satisfy their biological, sociological, and psychological needs – in other words, to be as fully human as possible.

When people recognize they have intrinsic worth, they are said to have self-esteem. Stanley Coopersmith concludes, on the basis of extensive research, that parents of children with high self-esteem:
(a) accept the child in his or her own right,
(b) lay down clear and enforceable rules of conduct, and
(c) allow the child a wide latitude to explore within those boundaries.
Firm and fair rules provide a secure world that makes sense, and freedom to explore it provides children with the confidence that they can make sense of it.

(f) *The person is not an interchangeable part*

If the person has intrinsic worth, then the person is not an interchangeable part. The shift from extrinsic to intrinsic motivation implies not only a shift from other-control to self-control but a shift from being other-employed to being self-employed. There are some encouraging signs that we can swell the ranks of the self-employed, that the work of the world can be done without stretching and chopping all of us to fit the Procrustean bed of jobs.

We have developed mechanical slaves to do mechanical jobs. Instead of being apprehensive about being replaced by a machine, we should be delighted. Machines, by definition, can only do mechanical things. Therefore, anyone who could be replaced by a machine *should* be replaced by a machine. The machine frees the person from mechanical tasks and thus sets the person free for human tasks.

The most encouraging signs, however, are not so much technological innovations but changes in cultural attitudes. Serious discussion of guaranteed annual income and negative income tax implies a shift from the traditional view that there are those who work and those who are supported by their charity, to the view that everyone in an affluent society should have enough to lead a simple, dignified life and that malcontents who aspire to luxuries are free to work to acquire them.

There is a growing counterculture of people who are content with a simple life and are not embarrassed about being 'unemployed' or about 'getting something for nothing'. They consider themselves as self-employed and consider self-actualization as a legitimate form of self-employment. They do not consider the traditional distinction between work and play very useful. One person's work is another person's play. One person can work at taking children to a fun fair whereas another person can play at solving quadratic equations. They prefer a distinction with respect to the subjective map rather than the objective world. The distinction between doing what you must do and what you want to do is such a subjective substitute for the objective distinction between work and play. Their criterion of success is the average number of hours a day they are doing what they want to do. If you are healthy what you want to do is to satisfy your organic biological, sociological, and psychological needs, to realize your full human potential, to self-actualize.

(g) *The person has intimate relationships*

If the person is not an interchangeable part, then the person has intimate relationships. All relationships are potentially intimate. A stranger is just a friend you have not met yet.

The contract (or, rather, the understanding) with your grocer is that you tacitly agree not to realize your potential intimacy. You will limit yourselves to exchanging money and groceries. After all, you can handle only so much intimacy – even if only for the simple fact that you have only so much time. There is however a penumbra of intimacy around your contractual arrangements. If your grocer falls off his stool while reaching for your cornflakes, then you go to his aid. If you are out of work and thus out of money, the grocer may extend credit until you are back on your financial feet again. Neither of you says 'That's not in the contract.'

Whereas the contractual relationship is based on the rules of human beings, the intimate relationship is based on the laws of nature. We recognize other people as members of the same species on the same planet in essentially the same predicament as ourselves. If God is dead, then there is no one here but us. Other people are the only personal element in an impersonal universe. They hold out the only hope of empathy, of understanding, of caring.[12]

Once again, the relationship between the person and the society is a macrocosm of the relationship between one person and another. The relationship is synergetic. That is, what is good for the person is good for the society, and what is good for the society is good for the person. Ruth Benedict puzzled for years while working as an anthropologist in various societies about the essential difference between the societies she liked and the societies she did not like. She finally concluded that, in the societies she liked, the ends of the society and the ends of the person were synergetic, and, in the societies she did not like, the ends of the society and the ends of the person were antagonistic. She gave her only copy of her notes on this synergetic-antagonistic distinction to Abraham Maslow, who used it in his consideration of our two basic problems – that of the good person and that of the good society.

Our prevailing philosophies of the relationship between the person and society (whether they be the bad-person, good society view of Hobbes or the good-person, bad-society view of Rousseau) see them as antagonistic. There is no reason, however, why the ends of the person and of the society can not be synergetic in our society. Society is a social invention and we may as well invent a good society. The good society is one which provides the commodities to satisfy the organic needs of the person, and the good person is one who has his or her organic needs satisfied. The good person is created by the good society, and the good society is composed of good people.[13]

Social psychologists have tended to focus on antisocial behaviour. It is more urgent and more dramatic. Some social psychologists have, however, begun to study prosocial behaviour – caring, sharing, helping, and other positive acts – that is, the behaviours that make a synergetic society possible. Paul Mussen and Nancy Eisenberg-Berg have summarized the research so far on the development of such prosocial behaviour in children.

Prosocial behaviour in children is increased by adults who:
(i) engage in prosocial behaviours (but not by those who demand prosocial behaviours – do-as-I-say-not-as-I-do does not work);
(ii) reason with them as a means of discipline;
(iii) encourage them to reflect on the feelings and expectations of themselves and others;
(iv) assign them early responsibilities for others (like teaching younger children);
(v) reward them for prosocial behaviour;
(vi) provide them with role-playing and empathy-promoting exercises;
(vii) make explicit demands that they act maturely.

4 From micro to macro

The ends and means of human development, from the outside-in point of view of behaviourism, were presented in section 2(a) *The person has only extrinsic needs* and section 2(d) *The person is conditioned from the outside in*, respectively. The person does not have goals but, in a sense, the goals have the person. The goals are in the external environment rather than in the person. The person does not develop but, in a sense, is developed. The person is conditioned from the outside in. This static model of the person reflects the fact that behaviourism is based on physics. As an infant science, concerned about out credentials, we modelled psychology on physics, the most respectable adult science. Thus, the person is viewed as an object 'which will remain at rest or continue to move in a straight line unless acted upon by some external force'.

The most charitable interpretation of behaviourism is that the goal is survival, which at least brings psychology within biology where it is obviously more at home than in physics. This is a worthy goal but a limited one, since it does not account for people who see success as somewhat beyond mere survival or for people who choose not to survive in undignified circumstances. Also it is still a static goal. My goal yesterday was to survive, today is to survive, and tomorrow is to continue to survive. The only dynamic quality in this model is the environment – that is, each of us must continue to survive under our changing circumstances.

If the goal of human development is survival, and the process of human development is conditioning, then the best indicator of human development is longevity. It does not matter what one does with one's life so long as it is long. A long miserable life is better than a short happy life. A person who lives an unexamined life for a century is better developed than Lord Byron who petered out before 30. Such inanities suggest that this model of the person is inadequate.

Unfortunately, this inadequate model is the implicit model of the person underlying the traditional practices in social development. I have tried to demonstrate this by going beyond the usual exposition of behaviourism, which considers only the above two propositions – 'The person has only extrinsic needs' under the topic of motivation and 'The person is conditioned from the outside in' under the topic of learning – to four further propositions which are implied by them.

The six propositions serve also to contrast the behaviouristic model clearly with the humanistic model, which is usually only vaguely described as a reaction to behaviourism. As you can see by looking at the table of contents, each of the propositions in the system of propositions, which constitutes the humanistic concept of the person, negates the corresponding proposition in the system of propositions, which constitutes the behaviouristic concept of the person.

The ends and means of human development, from the inside-out point of view of humanism, were presented in section 3(a) *The person has intrinsic needs* and in section 3(c) *The person grows from the inside out*, respectively. In this case, the goals are not in the environment but are built into the person at the moment of conception. The person is not manipulated from the outside in but unfolds from the inside out. Powered by an organic system of biological, sociological, and psychological needs, the person seeks satisfiers of them in order to fully realize the human potential. Those needs are universal within our species, providing us all with our common humanity, but the means of satisfying them vary from time to time and from place to place, providing us each with our delightful diversity. People are not passively 'developed' by forces outside their skin, awareness, and control, but are the active agents of their own development. It is not mere quantity of life but quality of life which matters. May I suggest that this alternative model of human development is a more adequate foundation for a theory of social development?

The extension of this inside-out point of view from the micro to the macro level is best left to those of you who are more experienced than I at the macro level of analysis. However, before leaving the subject, let me just make one distinction between the two points of view which may be of some help in this process.

The inside-out point of view of human development is based on a faith in biology, whereas the outside-in point of view is based on a faith in technology. The reason my chapter is subtitled 'or (better) on not turning development outside-in in the first place' is that I believe industrialized societies have replaced faith in biology with faith in technology. We have much faith in biology for the first nine months of human development (most of us with the possible exception of first-time fathers accept the 'miracle' of birth with surprising equanimity). However, we then lose the faith, and assume that the child will not grow up right unless we poke and prod it around from the outside with our various inventions – machines and institutions.

Joseph Pearce documents the damage done by our attempt to replace a faith in biology with a faith in technology with respect to child-rearing. Nature's plan for our children is that they evolve from the inside out through a series of known matrices which serves as secure bases for exploration of the unknown. Each transition from matrix to matrix – from womb to mother, from mother to world, from world to body, from body to mind – should be gentle, providing an optimal balance between the known-familiar and the unknown-unfamiliar.

The first transition from womb to mother – the process of birth – provides a dramatic illustration of the undermining of the predetermined inside-out plan by our outside-in technological interference. Mothers in traditional societies, without the dubious advantage of technology, provide their child with a gentle transition. Marcelle Gerber, with a research grant from the United Nations Children's Fund, discovered that children in Kenya and Uganda were calm, happy, alert and enormously intelligent. By contrast, birth in industrialized countries is a technological, profit-making event, with little respect for the integrity of the child and the participation of the mother, which lights the fuse of a psychic time bomb with untold subsequent damage.

This faith in biology extends beyond physical development to moral development. Most of us know, from our intimate sample of one, that we are basically good with superficial flaws and, through our capacity for empathy, that other people are basically good too. Kenneth Clark argues that most of our personal and social problems stem from the blocking of this natural empathy by culturally-acquired power drives.

Much recent evidence has accumulated to justify our faith in biology and to question the substitute false faith in technology. Every revelation about the wonders of the human body and mind is seen, by the simple-minded, as a point in favour of technology. We congratulate ourselves on discovering it, which is very understandable, but fail to go on to admire what we have discovered. It is as if I thought that my finding the Empire State Building was somehow more of a feat than the building of it. On deeper reflection, we realize that every discovery is only one small step for technology and one huge leap for biology.

Can this faith be extended from the micro to the macro level of analysis? We tend to think not. As we move into the rarified macro zone, we shift from biology to psychology to sociology and must replace any faith we may have in biological processes with a faith in cultural processes. We can, however, keep the faith. Capitalism, it is claimed, is partly based on an extension of the biological metaphor. Darwin's survival of the fittest principle was applied to the marketplace by the advocates of social Darwinism. However, more careful consideration suggests that this was more rationalization than rationale. It is more likely that our survival depended on co-operation rather than competition. It was the fact that we could work together in groups which enabled us to catch the faster animals and kill the stronger ones, and thus crowned this slow puny creature as King of the Jungle. Our genes may, indeed, have been fighting among themselves, as argued by the sociobiologists, but *we* were working together.

Notes

1 I am sorry about the neologisms. More words are not high in our priority of need. However, I cannot find any extant words which convey this important distinction. The closest I can come is the distinction within botany between endogenous (growing from within) and exogenous (growing by additions from the outside) factors in the growth of plants. It is ironic that the endogenous factor is recognized more in the case of simple biological systems like the yew and the iris than in the case of complex biological systems like you and I.

   A footnote about footnotes. This distinction, as will be argued later, is based on the distinction between the inside-out view of the person (experience) and the outside-in view of the nervous system (behaviour). I feel justified therefore in including some asides from my own personal experience as footnotes, which add some subjective flavour to the objective content in the text. This may seem somewhat self-indulgent but my own experience is the only experience I have had.

2 I anticipate – and welcome – some static from some readers for putting capitalism, socialism, and communism all in the same bed. They look like unlikely bed-mates. However, whereas communism and socialism have the redeeming grace of being inside-out in theory, they have tended so far to be outside-in in practice.

3 I owe this lovely image to my colleague Iris Fitzpatrick-Martin, but I hereby absolve her of any blame for the use to which it was put.

4 No doubt those who have a vested interest in the current view of development will try to head us off at Galtung Pass by inventing some nasty name for someone who attempts to discuss matters assigned to the molecular level at a molar level, just as they invented 'reductionist' to chide those who discuss matters assigned to the molar level at a molecular level. Perhaps we could prepare for this ambush by pre-inventing the term and rehearsing the arguments against it. Any suggestions?

5 When we refer to a person, we refer, more precisely, to the nervous system of a person. Our uniqueness lies in our nervous system. We talk to the eyes of the person rather than to the ears or the elbows because that is the only place where the nervous system is exposed. A brain transplant would differ dramatically from a heart transplant – it is the donor rather than the recipient who would survive in the recipient's body. 'Extrinsic' thus means 'from outside the nervous system' – that is, from the internal environment, consisting of the other subsystems of the body, as well as from the external environment.

6 I once had the following conversation with an immigration officer. 'How will you support yourself in the United States?' 'My publisher will pay me.' 'They can't do that.' 'Why not?' 'You will be depriving a United States citizen of a job.' 'They can't hire anyone else to write *my* book.' 'A good point. Have a good trip.' Few modern jobs offer such a satisfactory means of clinching an argument.

7 Another conversation – this one at a cocktail party for advertising executives. 'Who are you with?' 'I came alone.' 'I mean, which advertising agency do you work for?' 'Sorry – I'm not with any company of that kind either.' 'What do you do?' 'I do all sorts of things.' 'I mean, what is your job?' He didn't get around to asking me 'What are you worth?' but the look of disdain as he excused himself to seek more congenial company implied that he had already decided I was not worth much.

8 My hierarchy is a simplification of Maslow's hierarchy in which needs are clumped into three broad categories – biological, sociological, and psychological. His highest need – self-actualization – is considered here as the realization of the full human potential, which involves satisfaction of all three sets of needs.

Once again, I anticipate some static from readers who justly question the hierarchy of needs concept. I think, however, that some of the dissatisfaction with the concept may be due to the behaviouristic emphasis on biological needs, with the phenomena accounted for here under organically-based sociological and psychological needs viewed as mere cultural habits which can be ignored until the 'basic' biological needs are satisfied.

9 A fine conversation I was having at a party was interrupted by a third person. The quality of the conversation shifted suddenly in some strange way which I did not understand at the time. I now realize that we had shifted down the hierarchy of needs from the psychological to the sociological level. We had been playing with ideas but had been shifted to working on intepersonal problems.

We all know students who are so preoccupied with satisfying their sociological needs that they have little computer time left for the satisfaction of psychological needs – the luxury of knowing and understanding the world.

10 It is this free-will option which explains why I did not cast my inside-out outside-in debate within the framework of the nature-nurture controversy. That debate is concerned with the extent to which we are *determined* by genetic or environmental factors. Time after pompous time, psychologists (including one of my former selves) intone that 'this behaviour is determined by some complex interaction between environmental and genetic factors'. However, consider the Siamese twins, Chang and Eng. Leslie Fieldler reports that they had very different personalities – Chang was a drunk and a womanizer and Eng was a teetotaller and almost a celibate. Since genetic factors were identical and environmental factors were as close as possible for any two people, their profoundly different personalities must be attributable to some third factor. Could it be that Chang decided to lead a short happy life whereas Eng decided to try for a long modest one? Poor Eng had to die when Chang died. However, most of us are free to act on such decisions. The interactionism of Jean Piaget, which describes the optimal orchestration of inside-out genetic processes and outside-in environmental processes, suggests that human development is a process of gradual emancipation of a person from the tyranny of the environment.

11 This was pointed out to me by a wise old man I once met in Los Angeles. He had just emerged from a mental hospital, he was physically sick, his wife had left him, and his children were alienated from him. After a five-hour discussion, during which we disagreed on every topic that arose, I suddenly saw an underlying pattern to our debate.

'I see now, old man, why we disagree. We consider each topic at such depth that we get right down to our basic philosophical assumptions, and I am a free-willist whereas you are a determinist.'

'Of course, young man, you are a free-willist – your life is going well and you want to take the credit. And, of course, I am a determinist – my life is going badly and I don't want to take the blame.'

12 These two basic attitudes toward other people are nicely represented by two gestures I encountered while travelling in Nepal. The traditional gesture is to hold your hands as if in prayer, bow, and say 'Namaste', which means 'I honour the divinity in you'. The modern gesture – alas, in urban areas pervaded by the influence of Western industrialized nations – is to hold out one hand palm up and say 'rupee'. The shift from 'Namaste' to 'rupee' is symptomatic of the shift from intimate to contractual relationships, from relationships in a traditional society to those in an industrialized society.

13 I feel almost embarrassed to talk of the good person – it is so unfashion-

able. I once heard of a man who lived with a beautiful and talented actress and wondered why he was worthy of such a fine woman. He was pointed out to me in the street. He had a kind face but, no, he was not magnificently handsome. He was later introduced to me. His conversation was lively but, no, he was not brilliant. Even later he took me to his home. It was comfortable but, no, he did not appear to be fabulously wealthy. As I got to know him, I slowly realized that he was simply good. It is an interesting comment on me and my times that it took so long for me to consider that possibility.

References

Allport, G. W., 'The functional autonomy of motives', *American Journal of Psychology*, 1937, 50, p. 141-156.
Amin, S. *Accumulation on a world scale: A critique of the theory of underdevelopment*, translated by Brian Pearce. New York: Monthly Review Press, 1974.
Asch, S. E. 'Effects of group pressure upon the modification and distortion of judgments' in Maccoby, E. E., Newcomb, T. M., & Hartley E. L. (eds.) *Readings in social psychology* (third edition), New York: Holt, Rinehart & Winston, 1958.
Berelson, B. & Steiner, G. A., *Human behaviour: An inventory of scientific findings*, New York: Harcourt, Brace & World, 1964.
Bruner, J. S., *Toward a theory of instruction*, Cambridge, Mass: Harvard University Press, 1966.
Clark, K. R., 'Empathy: A neglected topic in psychological research,' *American Psychologist*, February 1980, pp. 187-190.
Coopersmith, S., *The antecedents of self-esteem*. San Francisco: Freeman, 1967.
Evans, C., *The mighty micro: The impact of the computer revolution*, London: Victor Gollancz, 1979.
Festinger, L., Reicken, H. W., & Schachter, S., *When prophecy fails*, Minneapolis: University of Minnesota Press, 1956.
Flavell, J. H., *Developmental psychology of Jean Piaget*, New York: Van Nostrand Reinhold, 1963.
Galtung, J., *The true worlds: A transnational perspective*, New York: The Free Press, 1980.
Gardiner, W. L., *The psychology of teaching*, Monterey, California: Brooks/Cole, 1980.
Gerber, M., 'The psycho-motor development of African children in the first year and the influence of maternal behavior', *Journal of Social Psychology*, 1958, 47, pp. 185-195.
Goldfried, M. R. & Merbaum, M. (eds.), *Behavior change through self-control*, New York: Holt, Rinehart & Winston, 1973.
Harlow, H. F., 'The nature of love,' *American Psychologist*, 1958, 13, pp. 673-685.
Heron, W., 'The pathology of boredom', *Scientific American*, 1957, 195(1), pp. 52-56.
Keys, A., Brozek, J., Henschel, A., Mickelson, O. & Taylor, H. L., *The biology of human starvation*, Minneapolis: University of Minnesota Press, 1950.
Kleitman, N., 'Patterns of dreaming', *Scientific American*, 1960, 203(5), pp. 82-88.
Lederer, K. (eds.), *Human needs: A contribution to the current debate*, Cambridge, Mass.: Oelgeschlager, Gunn & Hain, 1980.

Maddi, S. R., 'The search for meaning' in Arnold, W. J. & Page, M. M. (eds.), *Nebraska Symposium on Motivation (Volume) 18)*, Lincoln: University of Nebraska Press, 1970.
Maddi, S. R. & Costa, P. T., *Humanism in personology: Allport, Maslow, and Murphy*, Chicago: Aldine, 1972.
Mahoney, M. J. & Thoresen, C. E., *Self-control: Power to the person*, Monterey, California: Brooks/Cole, 1974.
Mallmann, C. A., Nudler, O., and Max-Neef, M. A., 'Quality-of-life-oriented development and global social modelling', Bariloche, Argentina: Fundacion Bariloche, 1979.
Maslow, A. H., *Motivation and personality*, New York: Harper & Row, 1954.
Maslow, A. H., 'The need to know and the fear of knowing', *The Journal of General Psychology*, 1963, 68, 111–125.
Maslow, A. H., 'Synergy in the society and in the individual', *Journal of Individual Psychology*, 1964, 20, 153–164.
Maslow, A. H., *Toward a psychology of being* (second edition), Princeton: Van Nostrand, 1968.
Milgram, S., 'Behavioural studies of obedience', *Journal of Abnormal and Social Psychology*, 1963, 67(4), pp. 371–378.
Mussen, P. & Eisenberg-Berg, N., *Roots of caring, sharing and helping: The development of prosocial behavior in children*, San Francisco: Freeman, 1977.
Nudler, T., 'Towards a model of human growth', *Project on Goals, Process and Indicators of Development (GPID)*, Tokyo: The United Nations University, 1979.
Pearce, J. C., *Magical child: Rediscovering nature's plan for our children*, New York: Bantam, 1977.
Richards, A., *Hunger and work in a savage tribe*. Reported in M. Bates. *Gluttons and libertines: Human problems of being natural*, New York: Random House, 1958.
Simon, H. A. *Sciences of the artificial*, Cambridge, Mass.: MIT Press, 1969.
Skinner, B. F., *Verbal behavior*, New York: Appleton-Century-Crofts, 1957.
Slater, P., *Earthwalk*, New York: Doubleday, 1974.
Spitz, R. A., 'Hospitalism: An inquiry into the genesis of psychiatric conditions in early childhood in Fenichel O. *et al.* (eds.), *The psychoanalytical study of the child (Volume 1)*, New York: International University Press, 1945.
Toffler, A., *The Third Wave*, New York: William Morrow, 1980.
Valaskakis, K., 'Eclectics: Elements of a transdisciplinary methodology for future studies' in CIBA foundation symposium 36, *The future as an academic discipline*, Amsterdam: Elsevier, 1975, pp. 121–143.
Valaskakis, K., Sindell, P. S., Smith, J. G., & Fitzpatrick-Martin, I., *The conserver society: A workable alternative for the future*, New York: Harper & Row, 1979.
Valaskakis, K., 'The information society: The issue and the choices, *Information Society Project*, Integrating Report, Montreal: GAMMA, 1979.
Valaskakis, K. & Sindell, P. S., 'Industrial strategy and the information economy: Toward a game plan for Canada, *Information Society Project*, Paper No. 1-10. Montreal: GAMMA, 1980.
Wilson, E. O., *Sociobiology: The new synthesis*, Cambridge, Mass.: Harvard University Press, 1975.
Wolfe, J. B. 'Effectiveness of token-rewards for chimpanzees', *Comparative Psychology Monogram*, 12, No. 60, 1936.

Section III

Human Development: Psychosocial Aspects

5
Classification and Hierarchization in Motivational Fields: Co-Evolution Vectors

C. Mamali and G. Paun

People satisfy and develop their needs by and through their relations with others. Needs satisfaction under conditions of isolation is an exceptional situation – even needs whose satisfaction has a strictly individual character (e.g. the need for sleep) are conditioned by the social relations in which the individual is actually living.

There is a close dependence between satisfaction, genesis and development of human needs, on the one hand, and the complexity of social relations evolution, on the other hand.

In the social interaction process, particular motivational relations also develop between the interacting parties. Motivational relations are especially defined by the quality and intensity of needs that are satisfied and generated within the framework of social interaction. The present study aims at elaborating a model of *classification* and *hierarchization* of the motivational relations occurring between the interacting parties in the process of needs satisfaction, with a view to specifying the *motivational field* and the *co-evolution vectors*.

What is the evolutive significance of the motivational relations between individuals and human groups? What are those motivational relations in which the satisfaction of the needs of individuals, groups or communities may favour their co-development? What are the main types of motivational relations and how could they be classified? How could a human group be characterized according to the existing motivational relations? To what extent and especially in what way could the social context and particularly the immediate motivational field influence the motivational relations between individuals or groups?

These are only a few questions which the present study seeks answers to, taking a holistic view.

These questions, which essentially concern the relation between the motivational development of an individual (or group) on the one hand, and the quality of the motivational fields he lives in on the other, are significant for the study of the human growth problematique. More specifically, human growth at individual, group and organizational levels, is not independent of the social fields in which they live. Part of the forces of these fields are motivational forces.

It is a matter of common knowledge that the surveys focusing on understanding of the stages of human development, either as moral development (J. Piaget, 1932; L. Kohlberg, 1972), the levels of interpersonal maturation (C.E. Sullivan *et al.*, 1957), the evolution of cognitive structures within social interaction (A.N. Clermont-Perret, 1980) the theory of satellization (D.P. Ausubel, 1952) or the theory of life-cycles (E.H. Erikson, 1963), attach varied importance to interaction in the social context in which development occurs, even if these variables are not always explicitly analyzed. Moreover, G.W. Allport (1937) showed that the development of personality involves deep-going changes in motivational structures, while more recent surveys (J. Nuttin, 1980; W. Lens and A. Gailly, 1978) using the Motivational Induction Method, show the changes according to age in motivational goals and future time perspective.

Human needs and social interaction

A varying number of social needs is included in most classical models of human needs. They may be either protection or parental propensity, submission or gregariousness in McDougall's system (1952) or recognition and affiliation needs in H.A. Murray's system (1938): anyway they imply a form of social interaction. In A. Maslow's hierarchy model of needs (1954) there are also some needs that are based on the individual's social interaction ('esteem' need). In the matrix conceived by C.A. Mallmann (1976) besides psychosomatic (intra-human) and environment (psychohabitational) needs, interhuman needs are also included.

Paradoxically, although these models include social needs, they do not go further than the individual level, thus overlooking the needs transformation in relation to the quality of interhuman relations. Irrespective of the size of the social needs set included, these types of models do not provide an image of the existing connections between the genesis and needs satisfaction and the dynamics of social relations.

Within Maslow's hierarchy, this disjunction has led to an atomistic outlook:

... in which the individual is perceived to be isolated and empty unless somehow fulfilled and actualized. In fact, this is a great problem of Western cultures, in which the individual is cut off not only from nature but also from other individuals (M. Maruyama, 1973, p. 55).

In order to avoid the pitfalls of such an approach, it is necessary to shed more light on the relationship between *social interaction* (especially interhuman relations) and *needs*. In the process of genesis, satisfaction and evolution of needs, the interhuman relations may be considered as: (a) evaluating criterion of needs; (b) means of generating, satisfying and transforming needs; (c) means of needs humanization; (d) interhuman relations represent in themselves a specifically human need.

(a) Interhuman relations – evaluation criterion of human needs. A tendency is accepted as a human need only by evaluating the direct or indirect consequences which the satisfaction of such a need may bear on the others, namely its consequences at interhuman level. This characteristic is used by J. Galtung in the definition of needs:

> . . . we shall not accept as a need anything that for logical if not for empirical reasons cannot be satisfied for all human beings. Clearly not every body can dominate, so the 'wish to dominate' does not qualify here as a *need* (the same, incidentally, applies to 'the wish to be dominated'). Correspondingly, a 'wish to be healthy' or a 'wish to be educated' would qualify, but not a 'wish to be more healthy' or a 'wish to be more educated than anybody else' (J. Galtung, 1978, p. 5).

The negative effects of meeting one's personal drives for the other parties become restrictive conditions in the determination of the human needs set.[1]

(b) Interhuman relations are a means of satisfying both the social needs and intra-human needs (in C. Mallmann's 1976 use of the term). At the same time, interhuman relations, by their complexity, may generate new needs and the transformation of existing ones.

(c) Interhuman relations by their character may determine needs humanization. Needs humanization occurs through their enrichment and transformation of the modalities to satisfy them.

(d) Interhuman relations represent in themselves a *specifically human need*. The identity needs, whose priority function is the avoidance of alienation (Galtung, 1978), are telling examples. However, we would like to emphasize the reciprocal character of the influences between the individual's motivational development level and the type of needs that are satisfied in his social relations.

The satisfaction of needs within the framework of interhuman relations has potentially a *reciprocal* character. Inter-personal actions are mutually reinforced:

> Social behavior is an exchange of goods, material goods but also non-material ones, such as the symbols of approval or prestige (Homans, 1967, p. 457).

The social exchange is real 'where determination is mutual' (Homans, p. 449). From this viewpoint, each participant in the interaction is potentially a source of material or non-material satisfiers and dissatisfiers for all those with whom he interacts.

The social exchange in which a variety of human needs is satisfied and generated simultaneously implied the *complementarity* and *reciprocity* of interhuman actions and reinforcements. A.W. Gouldner considers that reciprocity is a valid norm for all cultures and makes a clear-cut distinction between complementariness and reciprocity

> . . . complementarity connotes that one's rights are another's obligations, and vice versa. Reciprocity, however, connotes that each party has rights and duties (Gouldner, 1967, p. 277).

The reciprocity and complementarity principles have seldom been correlated with the problematique of human needs genesis and satisfaction, a fact that induced a non-dialectical view in conceiving the relations between them.

R.A. Hinde built a very useful outlook on this important topic considering that 'the interactions within relationships may show complex patterns of reciprocity and complementarity, with idiosyncratic patterns of imbalance' (1979, p. 80).

Each participant in interaction potentially has a *double motivational status*; on the one hand, he is a possible *source* of satisfiers and dissatisfiers for the others and on the other hand he is a possible *receiver* (consumer) of the satisfiers and dissatisfiers emitted by the others.[2]

Table 5.1

			\multicolumn{4}{c}{EGO}			
			Emits		Receives	
			Satisfiers	Dissatisfiers	Satisfiers	Dissatisfiers
ALTER	Emits	Satisfiers	+ 1 +	− 2 +	+ 3 +	− 4 +
		Dissatisfiers	+ 5 −	− 6 −	+ 7 −	− 8 −
	Receives	Satisfiers	+ 9 +	− 10 +	+ 11 +	− 12 +
		Dissatisfiers	+ 13 −	− 14 −	+ 15 −	− 16 −

Without going into the details of these motivational relations, we should mention, so far, that they are complicated by the type and quality of the satisfiers and dissatisfiers exchanged within interaction. They may be either intrinsic or extrinsic and may induce the satisfaction or the dissatisfaction of more complex or simpler motives. One should not forget either that the interacting parties may fail to accurately perceive the satisfiers and dissatisfiers emitted by others.

Optimum motivational relations are those in which both interactants emit and receive satisfiers (1, 3, 9, 11). Such relations imply both complementariness and reciprocity. However, not only the achievement of an equitable ratio between the amounts given and received by each party is interesting, but also the way in which these reciprocity relations influence the evolution of the needs systems of the interacting individuals or groups.

What type of relations may occur between two persons or groups having different goals or needs? Although Maruyama (1972, 1973) does not explicitly follow the dynamics of the motivational interaction, he outlines a theoretical framework with an heuristic value for this problematique.

Maruyama describes five possible types of interaction between groups: (a) groups which co-exist with no, or very little, interaction between them – *separatism*; (b) groups benefiting from one another – *symbiosis*; (c) groups which gain what the others lose – *parasitism*; (d) one group which harms another group – *antibiosis*; (e) many groups harm one another – *mutual antibiosis* (Maruyama, 1972, p. 9).

It is interesting to notice that the table includes parasitism but does not take in the situation of conscious sacrifice, when a group loses deliberately in order to facilitate the gain for another group – namely altruism. Also the symbiotic and anti-biotic relations are determined only by the interaction between two parties, being isolated from a larger relational context. The symbiotic relation between two groups may turn into antibiotic when approached from the perspective of a third group (even if, for the time being, the relations of the two groups with the third one are not explicit).

These relations that may emerge between persons and groups are not structured only in relation to the *goals* of the actors but also in relation to the *needs* which each actor satisfies for himself in the interaction as well as in relation to the needs he can satisfy for the other parties of the interaction.[3]

The significance of the motivational relation between two persons or two groups is dependent on the social context. Namely, the evolutive sense of the motivational relation between two (or more) actors depends upon the effects of these relations on the motivational development of the other actors. The present study starts from the premise that the balance of imbalance accompanying the dynamics of human needs, homeostasis, self-regulation and development mechanisms, are to be found not only at individual but also at interhuman level. The satisfaction of the individual needs means not only modification of the intra-individual motivational tensions, but also the *transformation of the motivational field* characterizing the interhuman relations in a given context. How can the interhuman motivational field be described? What are the main significances attached to the empirical hierarchies of motives when needs satisfaction is analyzed within the framework of social relationships? We do not promise to provide answers to these questions; however, we shall attempt to construct a model which at least will be helpful in getting a more comprehensive perception of the problem raised by the evolutive value of interhuman relations at the micro and macro-social levels.

Miller's highly synthetical fundamental work *Living Systems*, which puts forth a generalized systemic theory of living systems from the cell to international organizations, shows that the growth curves of populations living under the same ecological circumstances are mutually dependent functions (Miller, 1978, p. 78).

Some authors have outlined the role of positive relations (mutual aid) in the process of evolution (Kroprotkine, 1910) putting forth the idea of a struggle not only for survival, but also for 'co-growth', as asserted by Siu (1957).

In view of these elements, we believe that the study of the various relations between two or more actors mutually satisfying similar and/or different needs should take into account the evolution (at motivational level) of each interacting party, as well as the evolution of the relations between them. The analyses of the interdependencies between the evolutions of the various interacting actors satisfying varied needs led us to the working out of a new theoretical framework focusing on the concept of *co-evolution*. Co-evolution means inter-dependence between the evolutions of the various interacting social actors, so that the changes which occur in an actor's evolutional process are associated with the changes that occur in the evolutional process of the other actors with whom he interacts. Co-evolution is a dialectical process which has two opposing directions marked by a pole of *mutual involution* of the interacting actors (whereby one's involution is both an outcome and a condition of the other's involution) and a pole *of co-development* (or *co-evolution* proper).

Motivational balance

The motivational balance defines a motivational relationship between two or more actors. Essentially, the motivational balance is characterized by the quality and intensity of the motives that each actor satisfies in the interaction as well as by the quality and intensity of motives that each actor satisfies for the other parties of the interaction.

The interaction of persons, groups, communities with homogeneous or heterogeneous needs systems is also characterized by the changes that occur in the interactant's needs system.[4]

The interaction of the needs systems of different persons calls for a re-appraisal of the human needs hierarchization problem. The hierarchization of human needs may be perceived from an axiological normative and empirical perspective (J. Galtung, 1978, p. 13). From the axiological viewpoint, the needs hierarchization:

> ... poses threats not only to cultural diversity but also to human diversity within cultures, and throughout any individual human being's life cycle;
> ... will serve as a *carte blanche* for the type of policies that might guarantee security and economic welfare, but at the expense of considerable amounts of alienation and repression ... (Galtung, 1978, p. 13; p. 14)

The construction of a static hierarchy of human needs considered to be universally valid irrespective of the cultural context and the socio-historical stage is a misleading enterprise.

So, are hierarchies of human needs just white elephants? They are not. Human needs vary from culture to culture and from one historical period to another, both as regards the contents, the way of satisfying them, their complexity and their position in a specific hierarchy. The conflicts of interests between social groups and individuals are sometimes generated by their tendencies to satisfy similar but differently hierarchized needs. Also, needs and values are closely correlated, the value of an object is considered to be the function of the number of needs it may satisfy (Bossel, 1976).

We consider that an important issue is that of the *consequences* which the *interaction* between individuals and human groups (with *homogeneous* and/or *heterogeneous*, similarly or *differently hierarchized* needs systems) may have on the *evolution* of the respective actors and of the relations between them. These consequences of interaction over the actors' motivational evolution may be specified by means of motivational balance.

The matrix of motivational relations may thus specify the main types of relations emerging between the two actors in the interaction process and precisely their evolutive sense in relation to the reference system itself. Within the interaction, each actor potentially has a *double motivational status*: each is a source of satisfaction of the other's needs and beneficiary of the satisfiers produced by the other. This double motivational status is an existential characteristic of the actors; the motivational balance between them is generated in relation to this status.

The matrix of motivational relations (the co-evolution matrix) takes into account the following characteristics:
— the dynamic relation between the hierarchical motivational structures that are specific to each interacting party;
— the hierarchical level of the needs which each actor satisfies for

himself and the level of needs he satisfies for the other actor;
— the interacting parties may differ both from the viewpoint of the *types of needs* they possess (having relatively homogeneous or heterogeneous structures), of *needs hierarchization* and of their *priority* needs at the moment in question;
— each interacting party may have its own scale for the hierarchization of its own or the other's needs;
— the needs of the interacting parties and the needs of interaction may be differently hierarchized by the participants in interaction and/or by various observers;
— each of the interacting parties is at a certain priority motivational level;
— the priority needs within a given interaction may be identical with, higher or lower than the priority level of the needs reached by each interacting party.

The easiest situation is when the actors have the same set of needs which they hierarchize similarly. This type is rare and if rigidly applied, may entail serious distortions when the individuals have a set of different and/or differently hierarchized needs.

Since a classification and a modelling of the motivational relations would be very difficult should all the above possibilities be included, we shall confine the present stage to the consequences of these relations only in keeping with each actor's scale of needs as he perceives and hierarchizes them.

The needs satisfied within or due to the interaction between two actors can be assessed according to the prevailing motivational level that was specific to each actor prior to his involvement into the relation in question. This comparison results in nine types of motivational relations (see the matrix of motivational relations) which account for the co-evolutionary patterns that may take place between two actors, be they individual or collective.

The nine situations point to four main states of the motivational balance, which are given by: (a) *motivational standstill*, where the interacting parties preserve the same priority motivational level they possess. This situation actually corresponds to interhuman homeostasis; (b) *unequal and even contradictory motivational development*, where the development of one party's needs system entails the standstill or even regression of the other party's needs system; (c) *mutual motivational involution*, when during interaction both parties pass from higher to lower needs; (d) *motivational co-evolution*, when both interacting parties pass to higher needs. The transition of each of the parties to a higher motivational level becomes a catalyst for the motivational evolution of the other party.

Seen in time, the balance states practically become *stages* in the *co-evolution* of the motivational relations.[5]

CLASSIFICATION AND HIERARCHIZATION IN MOTIVATIONAL FIELDS

Table 5.2 *Matrix of motivational relations*

Place of needs determining A within interaction with B, as against his dominant motivational level (A)

B \ A	Interaction needs lower than its motivational level (−)	Interaction needs identical with its motivational level (=)	Interaction needs higher than its motivational level (+)
Interaction needs are lower than its motivational level (−)	mutual motivational regression I	one-sided motivational regression II	contradictory motivational relations (regression-development) III
Interaction needs identical with its motivational level (=)	one-sided motivational regression IV	balance V	one-sided motivational development VI
Interaction needs higher than its motivational level (+)	contradictory motivational relation (regression-development) VII	one-sided motivational development VIII	motivational co-evolution IX

Place of needs determining B within interaction with A, as against his dominant motivational level (B)

(A and B can be individuals, groups, human communities, etc.)

The nine basic types of motivational relations are exemplified by interaction between two actors. The significance of the relation between two actors is affected by the relations of the two actors with other individuals.

For instance, between A and B, co-evolutional relations may be established, as well as between K and L. If A and B have at the same time co-evolutional relations with Y, while K and L have reciprocal regressive relations with X, the outcome is a qualitative difference between the two diades (A–B and K–L respectively). In this event, the motivational balance between two actors depends not only on the relations between the two actors (within the diade), but also on the relations between these actors and the others. The integral evolutional value of the motivational balance may be evaluated only in relation to the vectors it generates within the general motivational field. Therefore, the motivational field in which two or more actors are included is helpful, on the one hand in assessing the evolutional value of the motivational relations between the factors and, on the other hand, it is an *amplification reduction* or *compensation* factor of the motivational tensions between the actors. The consequences of a regressive reciprocal motivational relation in a motivational field dominated by regressive motivational relations will differ from the consequences of the same motivational relation in a motivational field dominated by co-evolutional relations.

The motivational fields may influence what H.H. Kelley (1979) called 'the transformation of motivation', the manner in which 'the person's interaction is responsive to patterned aspects of their interdependence, each one's behavior being governed not only by his/her own outcomes but by the other's outcomes as well' (p. 9). Taking into account this prospect, we may assume that a person's awareness of a certain status within the motivational balance may trigger off his priority orientation to a certain motivational transformation.

The determination of the co-evolution vectors within motivational fields may facilitate a better understanding of some processes of human growth. T. Nudler (1979) believes that: 'not every small group . . . encourages or promotes the integral growth of its members . . . a human group is "healthy" when it promotes or encourages the maturity growth of all its members, and "sick" when it tends to atrophy it' (p. 23).

From this point of view, we believe that the diagnosis of motivational fields may open a way of facilitating the assessment of an individual's chance of personal growth within a certain group. In order to approach these psychosociological problems, a series of classifications is needed as regards motivational relations, personality types from the motivational viewpoint and the motivational field.

A model of the motivational field

By means of the motivational balance, we may characterize the relations between two actors; however, in more complex situations, the use of the balance is not sufficient. Within a given group, a person may interact with all the other members. Whether he is in motivational evolution or involution, whether he develops at the expense of the others or the others develop at his expense, that is, in general, the problem of the *diagnosis of the type of motivational relations* and it calls for the utilization of more powerful tools.

Further on we shall construct a model of the motivational field established at the group level, by using elements of the graph theory; within this framework, we suggest some ways of solving problems of the above mentioned type.[6] The resolutions we put forth involve computerization in the final phase (the Motiva program), so that real significant groups may be easily approached.

By a *motivational field* we understand a quadruple:

$C = (G, U, E, f)$

where G is a finite set of individuals (actors, subjects) we noted with $a_1, a_2, \ldots, a_n$;

U is a set of pairs (a_i, a_j), $1 \leqslant i, j \leqslant n$,

the presence of an (a_i, a_j) pair in U meaning that between a_i and a_j actors an interindividual relation is established;

$$E = \{+, -, =\}$$

f: $U \to E \times E$ is the evaluation function of each interindividual relation to the actors' needs level;

$f((a_i, a_j))$ is that pair of signs from E which characterizes the motivational balance of the a_i, a_j individuals.

A motivational field such as the above can be graphically represented as an oriented graph with the edges marked by means of the f function; at the ends of the (a_i, a_j) edge we write the signs specified by $f(a_i, a_j)$. For instance, let us consider the motivational field

$C = (G, U, E, f)$ where
$G = \{a_1, a_2, a_3, a_4, a_5\}$,
$U = \{(a_1, a_3), (a_3, a_1), (a_1, a_4), (a_4, a_1), (a_1, a_5), (a_5, a_1), (a_2, a_3),$
$(a_3, a_2), (a_2, a_5), (a_5, a_2), (a_4, a_5), (a_5, a_4)\}$
$E = \{+, -, =\}$,
$f((a_1, a_3)) = (=, +), f((a_1, a_4)) = (=, =), f((a_2, a_3)) = (-, -),$
$f((a_2, a_5)) = (-, +), f((a_1, a_5)) = (+, -), f((a_4, a_5)) = (-, +)$

The representation associated to this field is the following:

Fig. 5.1

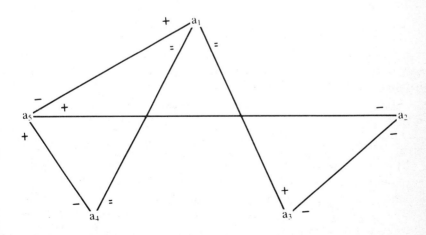

The writing of a field in symbolic form or its representation as a marked graph is the input information in the investigation aiming at diagnosing the actors and the group as a whole from the viewpoint of their motivational relations.

Let us first examine the maximum number of distinctive fields in relation to the number of existing relations, to be a group with R relations (R edges in the associated graph). Each relation is marked by two signs in E. The specification of R relations implies therefore the specification of 2R signs from E. There are 3^n distinctive strings of n length formed with signs from E, therefore there are 3^{2R} motivational distinctive fields with R relations. Further on, we give the values of these expressions for some R values:

Table 5.3

R	2R	3^{2R}
1	2	9
2	4	81
3	6	729
4	8	6561
5	10	59049
6	12	531441

Therefore, if it is possible to examine and diagnose the three relations fields, it is quite a problem to even generate a five-relation field. On the other hand, to have a conex $\delta = (G, U)$ graph, we should have:

card U $\geqslant$ card G−1

Let us now calculate the maximum number of distinctive fields according to the number of actors. We shall consider the groups with minimum and maximum distinctive relations. In a group with n actors, we have at least n−1 and at most $\frac{n(n-1)}{2}$ distinctive relations, therefore 2(n−1) and n(n−1) signs in E will identify the respective fields. In all, there are then $3^{2(n-1)}$, $3^{n(n-1)}$ distinctive motivational fields respectively.

The following table gives the values of these expressions for some values of n:

Table 5.4

n	2(n−1)	$3^{2(n-1)}$	n(n−1)	$3^{n(n-1)}$
1	0	1	0	1
2	2	9	2	9
3	4	81	6	729
4	6	729	12	531441
5	8	6561	20	3386784401
6	10	59049	30	....

It is true that in concrete cases the number of actors and/or of relations may be reduced before transition to the group diagnosis. For instance, not all relations are equally important for a given person; the belongingness of a person to a group is gradual, hence we could, in a first approximation, eliminate some actors, etc. Another solution would be to replace the relations between actors with relations between subgroups with particular cohesion.

Types of actors in a motivational field

Let $C = (G, U, E, f)$ be a motivational field with n actors. We may associate to each a_i actor in G, a nine-component vector $N(a_i) = (N^i_{..}, N^i_{+.}, N^i_{-.}, N^i_{.-}, N^i_{+-}, N^i_{.-}, N^i_{-+}, N^i_{.-}, N^i_{--})$, where N^i_{xy} is the number of motivational relations of a_i with other actors in the group; these relations are marked with xy.
Formally:

$$N^i_{xy} = \text{card}\{a_j \text{ in } G\ (a_i, a_j) \in U, f(a_i, a_j) = (x, y)\}$$

for each x, y from the E set.

Each component of the $N(a_i)$ vectors is comprised between 0 and n-1. On the basis of the value of these vectors, we may define motivational types of actors, namely imaginary ideal individuals with characteristic vectors strongly polarized in the directions specific to the respective type.

Let us try to define from the motivational point of view such personality types.

The altruistic type may be defined in several ways. For instance, he:
(a) increases the others' chances of satisfying their own needs, while reducing his chances of satisfying his own;
(b) increases the chances of the others' motivational evolution at the expense of his own motivational evolution;
(c) satisfies in interaction needs of a lower complexity level than that of his prevailing level, thus favouring the other parties' satisfaction of some of a higher complexity than that of their prevailing motivational level.

Therefore, the altruist type is characterized by a vector with higher $N_{-.}, N_{--}, N_{-+}$ components and smaller $N_{+.}, N_{.-}, N_{+-}$ components. We are less interested in the $N_{..}, N_{.-}, N_{.-}$; an intermediate value should probably be chosen for these components.

The egoistic type is complementary to *the altruistic type*. In his interrelations, this type:
(a) increases the chances of satisfying his own needs, reducing the other parties' chances to satisfy their needs;
(b) increases the chances of motivational evolution at the expense of the others' involution;

106 HUMAN DEVELOPMENT IN ITS SOCIAL CONTEXT

(c) satisfies, within interaction, needs of a higher complexity level than his predominant needs while the others satisfy needs of a lower complexity level than their prevailing motivational level.

In conclusion, the egoistic type has N_{+-}, N_{+-}, N_{--} maximum values, N_{-+}, N_{--}, N_{++} minimum values and intermediate values for the N_{--}, N_{++}, N_{--} components.

The altruist and the egoist are included in the more general *contradictory type*. The motivational vector of this type will then have N_{-+}, N_{++}, N_{--}, N_{--}, N_{++}, N_{++} maximum values and N_{--}, N_{++}, N_{--} minimum values.

The comparison between the characteristic vectors of the altruistic and egoistic types on the one hand and the contradictory type on the other hand, has led to the following conclusion: the contradictory type, six components were foreseen to reach a maximum value; it is clear, however, that it is more important that N_{-+}, N_{+-} be maximum and less important in the case of N_{--}, N_{++}, N_{--}. Thus the necessity arises for considering some marks of importance, some weights associated with the components of the motivational vectors. Each vector will be characterized by an M (type) and a W (type) vector.[7]

Let us now go further in our attempt at defining personality types from the standpoint of motivational relations.

The coevolutional type (the co-development oriented) corresponds to:
(a) the motivational situation in which satisfaction of one's own motives is a result and a prerequisite of satisfying the others' motives;
(b) the case when the transition of A from a specific needs level to a higher level is an outcome and a condition of the transition of A's interacting parties from a given to a higher needs level.

Therefore, we have N_{++}, N_{+-}, N_{++} maximum values (of higher importance being N_{++}), N_{--}, N_{--}, N_{--} minimum values (higher importance of N_{--}) and N_{--}, N_{+-}, N_{++} intermediate values, the N_{+-}, N_{-+} values being however smaller.

The reciprocal regressive type is defined by reversing the previous conditions:
(a) the dissatisfaction of one's own motives is an outcome and a condition of the dissatisfaction of the others' motives;
(b) the transition of a person from a given to a lower needs level is a condition and an outcome of the transition of the persons with whom relations are established from a given to an inferior needs level.

Therefore, we have N_{--}, N_{--}, N_{--} maximum values, N_{++}, N_{++}, N_{++} minimum values and N_{--}, N_{+-}, N_{-+} intermediate values.

The balanced type has N_{--}, N_{++}, N_{--} maximum values, N_{+-}, N_{-+} minimum values and N_{++}, N_{--}, N_{--} intermediate values and is defined by the conditions:

(a) the satisfaction (or dissatisfaction) of one's own priority motives is accompanied by the satisfaction (or dissatisfaction) of the others' priority motives;
(b) within the interaction, each person satisfies needs of the same level with his priority needs level.

In the same way, we could define *the unilateral evolutional type* (N_{+-}, N_{++} maximum, N_{++}, N_{--}, N_{--}, N_{--} minimum, N_{--}, N_{+-}, N_{+-} intermediate values), *unilateral involutional type* (N_{--}, N_{--} maximum, N_{++}, N_{--}, N_{--}, N_{--} minimum, N_{--}, N_{+-}, N_{+-} intermediate values) and other similar types.

An important aspect emerges. The previous considerations have a static character; they are made starting from a 'photograph' of the group at a given moment. If the respective group is examined at different consecutive moments in time, then the definition of the personality types from the perspective of motivational relations is a considerably more difficult task. For instance, the relation between two actors may be at each moment reciprocally regressive; however, when there is an *alternation in time* of the sense of this relation, it might happen that over a longer time span, the respective relation might happen to correspond to the actors' co-development. At each moment, one of them involves sacrificing himself to ensure the evolution of the other, but at the next stage, the motivational roles are reversed. For the whole period, both actors are gaining. The problem of the 'dynamization' of the previous model of the motivational field is very important and is worth being investigated in detail; we shall not dwell upon it here.

The definition of the characteristic M (type) and W (type) vectors is not a goal *per se*; by means of these vectors the actors of a motivational field may be hierarchized from the viewpoint of closeness to the considered types. For instance, the weighted distance from the N (altruism) vector to the vector describing a concrete individual from a given group represents the degree of altruism (in fact, the degree of *non-altruism*) of this actor. The arrangement of actors according to their degree of altruism provides the G hierarchy in relation to altruism. This type of hierarchies can be drawn up for all the personality types.

The strategy to follow in the hierarchization and classification of the actors of a motivational field may be closely followed when hierarchizing and classifying motivational fields. Field diagnosis can be made in this way, namely answers can be found to such questions as: Is the C field prevailingly altruistic or egoistic? Is it dominated by vectors of mutual regression or by vectors of co-development?

Such problems could be solved as soon as all the actors of the respective group have been hierarchized and classified. If the number of altruistic actors is higher than that of the egoistic actors, the respective field can be classified as altruistic. Further on we shall deal with a direct way of diagnosis. We have defined some types of actors; similarly, we may define *field types* and then study how a given case belongs to one type or another (Mamali and Paun, 1980).

At the field level the distinction between N_+^- and N_-^+ can no longer be made; such a relation is altruistic in *one sense* and egoistic in the other; it is perceived as a *contradictory* relation from outside the group.

Final remarks

In many respects, the present work has a preliminary character: many things which were only announced here are to be further investigated. A series of directions to follow have already been outlined. We mention again the necessity to modify the MOTIVA program to diagnose groups, the necessity to write programs for the classification of the actors in a group or of the groups, the practical check-up of characteristic vectors and of the weight vectors defining the *actors' types* and *groups' types*, the introduction of time in the model with all the ensuing consequences. The dynamization of the model is particularly important. As already mentioned, a contradictory relation alternating in time may correspond in fact, to an accentuated co-development process – a fact that is not grasped by the static approach to motivational balance.

The previous models may be used with a view to specifying the significant and insignificant actors. For instance, let C be a motivational field and an a_i actor. If, while ignoring a_i (and all relations that imply him), the typology of the remaining actors, the group type or other characteristics do not change, then the conclusion may be drawn that a_i is insignificant from the motivational standpoint. On the basis of a similar principle of invariance of the group characteristics the insignificant relations may also be eliminated.

An empirical problem which is worth approaching is a statistical study covering a number as large as possible of concrete groups in order to see what types of relations and actors, what triads and groups are more frequent, what interesting information is likely to be found on the motivational 'discharge' of actors: an actor who involutes in relation to some actors is likely to have relations in which he grows, thus compensating the loss from the first two relations.

Analysis of the motivational field and specification of the prevailing motivational relations and especially of the weight of the co-evolutional motivational vectors open an application field in societal learning which – as was demonstrated in the recent report to the Club of Rome – is not only innovative but also participative (J. Botkin, M. Elmandjra, M. Malitza, 1979). It is supposed that the chances of achieving societal learning will be higher in the predominantly co-evolutional motivational fields.

The wide spectrum of results that may be obtained by investigating motivational fields justifies the undertaking of further, deeper, more detailed studies.

In the end, let us advance a couple of hypotheses, which could also make up possible research directions, regarding the relation between the stage of the personality's moral development and the quality of the motivational fields in which he/she is involved.

The better the quality of motivational fields (i.e. the higher the number of co-development vectors), the greater a person's chances to pass to higher stages of moral development and the faster the transition.

On the other hand, the higher a person P's stage of moral development, the less disturbed his/her behaviour by a negative motivational field.

Notes

1 Kant's moral imperative which says that 'man is for man' but a goal is also true in the determination of human needs.
2 Within a social structure, individuals and human groups may be differentiated in relation to their role in producing the satisfiers and dissatisfiers.
3 Some authors consider the motive to be an element of synthesis between the need and the goal: A motive is a need for a desire together with the intention to reach an adequate goal (D. Krech, R.S. Crutchfield, N. Livson, 1969). Other authors consider that one and the same activity of more subjects oriented towards a single goal may spring from different motives; also an activity based on the same motive with more subjects may be directed towards different goals (M. Golu, 1975). In its turn, the relation motive-goal may be consonant or dissonant (A.N. Leontiev, 1959).
4 Even if we agree with the thesis according to which needs are localized in individuals, the fact that some needs are not satisfied may induce pathological processes not only at the individual but also at the group and community level. The dissatisfaction of a need may result in the disintegration of social groups (e.g. family, social class, etc.) while the individuals of these groups do not undergo the same process – at least for the time being.
5 In some cases, the 'level' may be replaced by the quality and intensity of the satisfied needs.
6 The psychological field theory was formulated by K. Lewin (1964), and was also called topological psychology. Q. Wright (1955) formulated a field theory of international relations. Rudolf Rummel (1965) developed a formalized field theory: 'the theory analytically divided social reality into two vector spaces. One space is that of attributes of social units, be they individuals, groups, or nations, and the other is that of the behaviour between social units. Within the attribute space, each social unit is located as a vector in terms of its attributes. Within the behaviour space, every pair of social units, called a dyad, is located as a vector in accordance with the interaction of the two members' (page 184). See also L.R. Alschuler (1973).

7 The Motiva program allows for the introduction of these vectors according to the wish of the users, therefore ensuring a good elasticity from this point of view (Mamali, Paun, 1980).

References

Allport, G.W., *Pattern and Growth in Personality*, New York: Holt Rinehart & Winston, 1937.
Alschuler, L.W., 'Status equilibration, reference groups, and social fields', in *General Systems*, Vol. XVIII, pp. 99–118, 1973.
Ausubel, D.P., *Ego development and personality disorders*, New York: Grune and Stratton, 1952.
Bossel, H., 'Notes on Basic Needs, Priorities and Normative Change' in Bisagno P. and Forti A. (eds.) *Research and Human Needs*, UNESCO, C.H.R., 1976.
Botkin, J., Elmandjra, M. and Malitza, M., *No Limits to Learning, Bridging the Human Gap*, Oxford: Pergamon Press, 1979.
Calude, C., and Paun, G., *Modelul matematic - instrument si punct de vedere*, Editura ştiinţifica şi enciclopedica, Bucureşti, 1980.
Erikson, A., *Childhood and society* (2nd ed), New York: Norton, 1963.
Galtung, J., *Methodology and Ideology. Theory and Methods of Social Research*, Vol .I, Copenhagen: Christian Ejler, 1977.
Galtung, J., *The Basic Needs Approach*, GPID/UNU Project, 1978.
Golu, M., *Principii de psihologie cibernetica*, Editura ştiinţifica si enciclopedica, Bucureşti, 1975.
Golu, P., *Motivatia - un concept de baza in psihologie*, Revista de psihologie, 1973, 19, 3.
Gouldner, A.W., 'The Norm of Reciprocity - A Preliminary Statement' in Hollander E.P. and Hunt R.G. (eds.) *Current Perspectives in Social Psychology*, New York: Oxford University Press, 1967.
Healey, P., *Basic Human Needs. Methodology and Mobilization*, GPID/UNU Project, 1979.
Hinde, R.A., *Towards Understanding Relationships*, London: Academic Press, 1979.
Homans, G.C., *Social Behaviour as Exchange* in Hollander and Hunt, op cit., 1967.
Jantsch, E. *Design for Evolution, Self-Organizing and Planning in the Life of Human Systems*, New York: George Braziller, 1975.
Kelley, H.H. *Personal relationships. Their structures and processes*. Hilsadale, New Jersey: Lawrence Erlbaum Associates, 1979.
Kohlberg, L. and Gilligan, C., 'The adolescent as a philosopher. The discovery of the self in a Postconventional world' in Kogan, J. and Coles, R. (eds.) *Twelve to sixteen: early adolescence*, New York: W.W. Norton, 1972.

Krech, D., Crutschfield, R.S. and Livson, M., *Elements of Psychology*, New York: Alfred a Knopf, 1969.
Kropotkine, P., *L'Entre aide, Un facteur de l'évolution*, Paris: Libraire Hachette, 1910.
Lederer, K., *Reflections about needs*, GPID/UNU Project, 1978.
Lens, W. and Gailly A., 'Content and future time perspective of motivational goals in different age groups' in *Psychological Reports*, University of Leuven, number 10, 1978.
Leontief, A.M., *Probleme ale dezvoltarii psihicului*, Editura ştiinţifica, Bucureşti, 1959.
Lerman, S.C., *Les bases de la classification authomatique*, Paris: Gauthier-Villars, 1970.
Lewin, K., *Field Theory in Social Science*, Cartwright D. (ed.), New York: Harper Torchbooks, 1964.
Mallmann, C.A., 'The Quality of Life and Development Alternatives' in Bisogno and Forti, op. cit., 1976.
Mallmann, C.A. and Marcus, S., *Empirical Information and Theoretical Constructs in the Study of Needs* (I), GPID/UNU Project, 1979.
Mamali, C., *Balanţa motivaţionala şi coevoluţie*, Revista de psihologie, 1979, N° 3.
Mamali, C. and Paun, G., *Classification and hierarchization in motivational fields: co-evolution vectors*, GPID/UNU Project, 1980.
Maruyama, M., 'Towards Human Futuristics' in *General Systems*, 1972, Vol. XVII.
Maruyama, M., *Paradigmatology and its Application to Cross-Disciplinary, Cross-Professional and Cross-Cultural Communication*, The IX Congress of Anthropology, Chicago, 1973.
Maslow, A.H., *Motivation and Personality*, New York: Harper & Row, 1954.
McClelland, D.C. and Winter, D.G., *Motivating Economic Achievement*, New York: Free Press, 1969.
McDougall, W., *The Energies of Men: A Study of the Fundamentals of Dynamic Psychology*, London: Methuen, 1932.
Miller, J.G., *Living Systems*, New York: MacGraw-Hill, 1978.
Murray, A.H., *Explorations in Personality*, New York: Oxford University Press, 1935.
Nudler, T., *Towards a model of human growth*, HSDR-GPID-7/UNU P-59, 1979.
Nuttin, J., *Motivation et perspectives d'avenir*, Presses Universitaires de Louvain, 1980.
Perret-Clermont, Anne-Nelly, *Social interaction and cognitive development in children*, European monographs in social psychology, No. 19, London: Academic Press, 1980.
Piaget, J., *Le judgement moral chez l'enfant*, Paris: P.U.F., 1932.
Rummel, R.J., 'A field theory of social action with application to

conflict within nations' in *General Systems*, 1965, Vol. X, pp. 183–211.

Siu, R.G.H., *The Tao of Science, An Essay on Western Knowledge and Eastern Wisdom*, Cambridge Mass: MIT Press, 1957.

Sullivan, C.E., Grant, M.Q. and Jr. Grant, 'Development of interpersonal maturity: application to delinquency' in *Psychiatry* 20, 1957.

Wright, Q. A., *A study of international relations*, Chicago: Univ. of Chicago Press, 1955.

6
Small Groups and Personal Growth: Distorting Mechanisms Versus the Healthy Group

Telma Barreiro

1 Incidence of small groups in the development of the individual 'self': the primordial primary group and other significant groups

The discovery of the incidence of others in the construction of the self can be considered without any doubt to be one of the main achievements of the human sciences.

To recognize that interaction with other human beings not only provides the context of the self but that it also intervenes in its very construction, is today an inevitable starting point for individual psychology.

The crucial importance of the original family group (Primordial Primary Group – PPG) in this construction process of the person and of the mother-child dyad within the group, can hardly be overrated. This has been generally recognized, as is well-known, by psychological research and clinical practice although perhaps starting from different paradigms. The deepest layers of personality, the basic patterns of our adaptation to reality, the fundamental frame of perception of others and self-perception, are constituted through the first experiences of communication and human interaction.

The intense light which psychological research has shed (and continues to shed) on the PPG has, in contrast, left the other human groups in which the individual develops his being in the shade (and even perhaps in the most absolute darkness).

While the incidence of the PPG on the personality is *ontogenetically decisive*, the nature of the other groups into which the individual must integrate, cannot be without importance for the future development of the personality. Just as everyone at birth is, without prior choice on his part, 'thrown' into a certain PPG which will leave a permanent mark on his existence (putting *a priori* conditions on all subsequent choices), so is he later, in most cases, also 'thrown' into other small but significant human groups for prolonged periods. He must adapt himself to these groups in order to survive 'socially', e.g. the school group, the work group. In these groups the individual can come to feel strongly committed, whether because of the assiduity with which he finds himself obliged to be present and interact with the other members, or of the social legitimacy which this membership means to him.[1]

Although it is a fact that the individual will go to all the non-primordial groups equipped with his patterns and the adaptation mechanisms he has built up during his first years of life within the PPG, it is also true that those new significant groups into which he must integrate will demand of him an effort to adapt which can lead him either to consolidate or ratify, or to modify those patterns more or less profoundly.

Let us consider, for example, the case of A who has had his first experiences in an aggressive, hostile family environment. His communication patterns have been making up a basic defensive/offensive scheme. The perception of his own identity is unsure and weak. In his efforts to keep up his defensive barriers, he will have few possibilities to continue without friction the construction of a non-dependent, creative self. Given this 'initial' definition of his adaptation mechanisms, it could happen that his experiences of belonging to new groups will or will not confirm his first experiences of aggression. If subsequently A should have the good fortune to enter a group which is very significant to him, where aggression is absent and therefore not an efficient mechanism, he would find himself somewhat defenceless in his mechanisms of adaptation to the environment and this would influence his personality to a certain extent. He would have to try to reconstruct his communication mechanisms in order to be able to remain in the group.

But if a structured mechanism of adaptation to human relations which the individual brings with him to his membership groups is to be modified as he passes through these groups, they should really not tolerate, but in fact actively reject, the previous mechanisms as an efficient form of adaptation. Otherwise it is very probable that the individual will not alter his basic patterns of interaction with the environment, due to the profound inertia of the structures acquired in the PPG.

In this sense, it happens that the groups may have latencies which an individual, comfortably set in his patterns, can arouse. This kind of 'vulnerability margin' of a group can allow the individual to cling to his consolidated structures, precipitating the latent tendencies in the group.

2 Distorting mechanisms of communication and human interaction

We have pointed out in 1 above the importance of the groups regarding the direction which a person's psychic development is to take.

We shall now mention certain distorting mechanisms of communication and human relations which can appear in the groups, with an adverse effect on individual personality. Such mechanisms, to the extent that they distort human interaction, negatively influence the formation and/or development of the self and actively hinder personal growth towards maturity. These distorting mechanisms can occur both in the primordial and in the non-primordial groups, although the cluster of mechanisms is likely to differ in one and the other. Both theoretical research and clinical practice have, from different conceptual angles, been revealing and analyzing various mechanisms of this kind, although centring their analysis almost exclusively on the primary relations within the PPG, where it is assumed that the fundamental origin of the structure of an infirm personality is to be found.

In this sense and as was pointed out in the preceding paragraph, it would be interesting to analyze what happens with those distorting mechanisms within non-primordial significant groups, such as working groups (in an office, workshop, etc.) school groups (both infants and adolescents) groups of professionals, academic circles, etc. There has not yet been any systematic exploration of how the individual personality, at a relatively profound psychic level may be affected by being submitted for many hours a day to the influence of these mechanisms, with few possibilities of change within a non-primordial group which could mean much to the individual as a vehicle of adaptation and survival within society, a group from which the individual could not escape.

Some of the most outstanding distorting mechanisms are the following:

(a) Authoritarianism. Domination by one or more over another or others. Authoritarianism can be open or concealed. It relies on fear of punishment, penalty, ridicule, loss of affection or protection, expulsion, etc. It can also rely on unconditional admiration.
(b) Instrumentalization. 'Use', assimilation of the person to a thing. Exploitation of one or more by another or others.
(c) Rivalry. Struggle between group members for money, prestige, power, the leader's affections, etc.
(d) Humiliation. Open or concealed. Degradation, irony, mockery.

Intolerance of mistakes and 'deviation' from the norm.
(e) Dual message. Dual language, double bind, distortion in communication.
(f) Pretence. Own defects, errors or conflicts deeply concealed; deceit, presentation of an 'image', taboos, etc.
(g) Aggression. Physical or psychic. Discharges of aggression, tension waves, etc.
(h) Elitism. Exaltation of some, disparagement of others. Applauding the 'winner', deriding the loser. Unfair distribution of goods between group members. Margination.
(i) Symbiosis. Dependent relationships; affective, intellectual and other dependencies. No clear discrimination of identities.
(j) Affective blackmail. Affection or gratification conditioned to compliance with rules or acceptance of authority.
(k) Denial of subjectivity. Denial of individual experience, denial of what belongs to each individual personality alone. Homogenization into a 'crystallized' objectivity.
(l) Distance. Indifference, incommunication, coldness. Undisguised lack of interest in the problems of others.
(m) Formalism. Ceremony, solemnity, emphasis on outward forms.
(n) Depression. Lack of passive stimuli, passiveness, boredom, necrophilia.
(o) Aggressive ethnocentrism. Perception of 'us' in opposition to the 'others'. Converging around aggression or defence.

These distorting mechanisms create or encourage *motivation lines*, helping to consolidate certain individual motivational systems. Some of them are used by the group leader to release motivations and encourage 'useful' behaviour. Thus, for example, members may work for fear of authority, or to compete successfully against another, or with the hope of passing from being 'instrumentalized' to being an 'instrumentalizer'. There is fear of aggression and many kinds of behaviour are resorted to in order to avoid it.

The dynamics of these mechanisms and the reasons they are efficacious in encouraging cohesion and the 'suitable' functioning of the group, relate to profound, psychic layers. They do not depend (at least in the majority of cases) on a conscious, 'Machiavellian' desire, but they are rooted in part in subconscious facets of the individual psychic life of the members and, in part, in the psychic history of the group itself as a structure. They are in turn inserted in a specific cultural frame.[2]

3 Effects of these mechanisms on the person

It is likely that only one of these mechanisms occurs, operating in isolation. Generally several of them appear together in a cluster, potentiating each other reciprocally. It would be necessary to analyze in complete detail what each of these mechanisms consists of and the deteriorating effect which each one (or perhaps each cluster) has upon individual personality. Such a task is beyond the limits of this chapter. We can only point out here very briefly some effects which in general these distorting group mechanisms have upon persons.

(a) They create a high degree of dependent identity. They stimulate anxiety and tension in the constant hankering after approval and acceptance.
(b) They undermine self-confidence, they do not stimulate self-esteem.
(c) They do not allow deepening or appraising of one's own experience (or that of another) and, as a result, they do not encourage originality and individual creativity.
(d) They force defensive/offensive barriers to be raised which hinder or impoverish interpersonal communication; they create resentment and distrust of others.
(e) They mutilate the sentiment of transcendence towards others, they encourage primary egocentrism.
(f) They drive towards a 'defective' growth. Instead of stimulating the development of aptitudes in function of personal growth or of active social solidarity, they lead to the development of the parts of self which will be 'efficient' or suitable for a 'good' adaptation.

In brief, we could say that these mechanisms basically affect self-perception, they impede development of a mature identity and they block (sometimes subtly though always effectively) human communication. These three deficiencies feed each other reciprocally, constituting a real vicious circle of generation of neuroses.

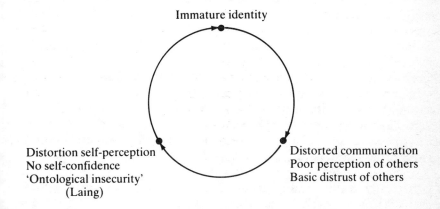

Perhaps these distorting mechanisms are at the basis of the anxiety and anguish which are characteristic of the human being of our times.

4 Cluster of mechanisms in the different groups

These mechanisms tend to become grouped differently, in accordance with the nature of the group. Some do not appear in certain types of groups and others, on the contrary, appear very frequently. This is not of course predictable with absolute certainty; there are only more or less probable combinations according to the nature and purpose of the group.

Thus, for example, in the primordial primary groups there can be: Authoritarianism (open or concealed); Rivalry (between brothers for their parents' affections; between husband and wife for power or prestige or for the children's affections); Dual messages, double bind; Affective blackmail (subordination of affection to compliance with norms); Symbiosis; Pretence (showing an image of security, firmness honourability, etc.; hiding conflicts and disappointments, taboos about sexual subjects etc.); Depression.

It is not so likely, for example, that Denial of Subjectivity, Distance, Formalism and Instrumentalization will occur.

In the working groups there could be: Authoritarianism, Instrumentalization, Rivalry, Aggression, Elitism, Denial of Subjectivity, Aggressive Ethnocentrism. Symbiosis or Affective Blackmail are less frequent.

In academic or artistic intellectual groups there tend to be: Authoritarianism (concealed); Elitism, Denial of Subjectivity, Formalism and, above all, Rivalry (for prestige, power and even economic rivalry). There can also be Aggressive Ethnocentrism (rejection of other research currents or other schools). Symbiosis or Depression are less likely to occur.

The greater the quantity, intensity and integration (mutual potentiation) of these mechanisms, the greater the power of the group to generate pathology in its members. We shall call these pathogenic groups 'infirm' groups. One group can be, as has been said, more infirm or less so and generate, unloosen or encourage therefore more sickness or less in individuals.

If within a given society these distorting mechanisms are repeated systematically in the small groups, we could say they constitute something like a basic psychosocial climate, a cultural feature. It is then much more difficult to overcome and it penetrates deeply into the personality. In this case it is very likely that the individual will acquire his basic forms of adaptation in a primordial group which displays some of these pathogenic mechanisms. And these adaptation mechanisms will subsequently be suitable for him to move within the general psycho-social environment which will finally help to reinforce them.

5 The healthy group

Contrariwise, the healthy group is one which creates mental health and encourages psychic maturity growth in its members.[3]

What is a healthy group? Within a healthy group there is a climate of acceptance, affection and real basic security. There exists an atmosphere of reciprocal sympathy, its members try to perceive, to really 'feel' the other, with his deepest needs, and to stimulate in him his human quality and individual potentialities.

A more analytical description of the features of a healthy group would be the following:

(a) There is no 'authoritarian' authority. The leader, if there is one, accepts or incorporates criticisms, is willing to give the grounds for his decisions, does not try to impose arbitrary rules, etc.

The leader is sensitive to the group's needs and possibilities and encourages psychological growth towads maturity in each of the members, so that his own role becomes increasingly less relevant.

The leader does not feel his leadership to be a form of power, he is even well-disposed to abandon his leadership completely in the event that the group matures and can dispense with its leader. In point of fact, in an optimum situation of mental health and maturity of members, the leader does not exist (or at most, leadership is by rotation).

(b) The entire group duly respects and values all of the members for what they are, not so much for their knowledge, skills or abilities as for their human condition and latent potentialities; no member feels slighted or marginated in his personal identity.

(c) A positive affection flows between the members of the group. They have faith and confidence in each other.

(d) Each member is encouraged to use his best endeavour to create and express himself within the group.

(e) The importance of subjectivity and individual experience is regained. It is accepted that every member has conflict, distress and suffering, hopes and needs. It is accepted that every member has his own particular personal history.

(f) Conflicts are brought out into the open without misgivings. The attitudes of members are discussed sincerely and openly. There is no hypocrisy or feigning. Members try to be as sincere as possible with each other and air their criticisms, fears, doubts as regards the rest.

(g) If necessary, if fundamental discrepancies arise, hidden underlying

values are explained. There is discussion and attempts are made to achieve coherence between values and to avoid 'double' standards. Differences of opinion and individual points of view are accepted.
(h) There is no rivalry, either for (i) power (since there is no irrational authority); (ii) for the respect and consideration of the others (since all the members are accepted for what they are); (iii) for 'prestige' (which in the end is the same as the preceding point); or (iv) for the affection of the authority (since every member finds his affective gratification within the group as a whole).
(i) Questions of substance have more appeal than formal, ceremonial or ritual questions.
(j) The members are not concerned with 'displaying an image'. They do not worry too much if others are aware of their supposed defects or shortcomings. They do not try to hide behind a mask in order to be more acceptable or worthy of respect.
(k) The group, as such, tries to transcend outwards to other groups or individuals. It is not a symbiotic, self-satisfied group. It does not generate ethnocentrism. It does not cause rivalry with other groups, but rather the contrary.
(l) Within the group there are no instrumentalization relationships between members (a member is not used as a 'tool' or a 'thing' by another). The group does not instrumentalize its members, nor do the members of the group instrumentalize each other.

Let us look at some of the effects of the healthy group on the individual personality.

If there is a truly egalitarian participation in the group, if leadership is authentically democratic and each person knows he will be consulted and his opinion appraised and heard with respect, his self-perception will be enhanced, he will regard himself with greater dignity and real self-esteem.

When he knows he will be well received, that there will be a profound respect for him, that his needs will be accepted and understood, he will not 'defend' himself or at least his defensive barriers will be weakened. To belong to such a group brings about a remarkable reduction in the level of anxiety over one's own identity. To the extent that each one feels there is no indirect or covert aggression, there is renewed hope of achieving a worthy, univocal, entire individual identity, recognized and accepted by the others. Very probably a sentiment of co-operation and solidarity with the group will be born in him through affection towards his companions and the pleasure he feels at being in their company. He will want to do things for the others because the individualist-egoist pattern will no longer be functional.

If the individual feels that his particular inner life, his personal experience and his personal history are not something dark and disposable but instead have value and meaning in interpersonal communication and that, in addition, he probably has points in common with the aspirations of other members of the group, the way he feels about himself and others will be humanized.

6 The psycho-therapeutical group

For many years now psychology has incorporated the group as a vehicle for certain forms of psychotherapy. Group psychotherapy recurs (explicitly or implicitly) to the fundamental assumption that what has been wrongly constructed in human interaction can be reconstructed in a healthier manner by also recurring to human interaction. In this sense, the basic philosophy of group psychotherapy bears out what we have been maintaining in this article on the role which a belonging to non-primordial groups can play in individual health (or illness).

In view of the enormous importance which the psychotherapeutic group has for the patient, it is particularly important that distorting mechanisms do not appear in it. In this aspect the therapist's work is crucial. The mechanisms will tend to appear inasmuch as they form part of the general psychosocial atmosphere. But it depends on the firmness of the leadership and the type of leadership employed whether they serve as 'efficient' adaptation mechanisms within the group or not. The fundamental danger here is that the therapist himself may find himself trapped somehow by the interplay of these mechanisms and may recur to them to handle the group. Here is where it becomes difficult to distinguish what is purely technical and professional from what enters into the ideological and axiological fields (the group leader's own system of values); and also what is related to the psychotherapist's own maturity or personal growth. The group leader may very well feel tempted to exercise an authoritarian control, to use irony or try to project an attractive image of himself. He may go so far as tolerating rivalry or even encouraging it. To the extent that the distorting mechanisms of human relations constitute a general psychosocial atmosphere, they can envelop even the therapists themselves like a fog. In the therapeutic group these mechanisms are doubly serious because they shut off the patient's possibilities of release. If, in the very place where a person is seeking a chance of mental health he encounters once again the mechanisms of communication and human relationships which keep him submerged in his identity immaturity and in his impossibility of profound human communication, this will exert a devastating effect on him.

7 The encounter group

In this sense, the so-called 'encounter' groups movement with its different varieties of groups is very interesting.[4]

These groups attempt to encourage an encounter between subjectivities, with a diminishing use of the 'mark' or image for the social presentation of the self and an increasing openness of one's own subjectivity to the subjectivity of the other. There is also an attempt to create in some way a non-authoritarian atmosphere where criticisms and negative feelings towards other group members or the leader can be expressed. Through the encounter each patient can find again his own underlying subjectivity and the aspects of his personality which he has been hiding or denying in order to obtain social consensus. He can even (according to the type of group involved) rediscover repressed aspects of his corporal sensitivity which he had been using or had simply maintained in a state of underdevelopment.

In relation to the central thesis we have been developing, these encounter groups would appear to be directed somehow to combat actively some of the distorting mechanisms which tend to exist in groups, such as: simulation, distance, formalism, denial of subjectivity and authoritarianism. And to combat these mechanisms, they recur to the potential healthy facets which there may be in the patients, to their latent capacity for positive experiences.

8 Is the healthy group a Utopia?

Perhaps this chapter will elicit the comment that the mechanisms we call here 'distorting' are inevitable (and even necessary) in groups and that the possibility of the existence of a healthy group is a Utopia. This sceptical attitude can arise from denying feasibility to the healthy groups in different senses: psychic, psychosocial, macrosocial or historical.

It is important to note that these forms of scepticism as regards the healthy groups differ one from the other. Some are more serious and definitive than others. The most radical is that which denies the psychic feasibility of the healthy group and could be expressed as follows: 'In view of the intrinsic nature of human psychic processes, these so-called "distorting" mechanisms will always appear, though strictly speaking they are not distorting, since in some way they are necessary resorts in the normal development of persons and in their social integration.'

As regards psychosocial feasibility, the sceptical stance can claim that the nature of all groups, of all human group interaction and communication presupposes the existence of some of these 'distorting' mechanisms. This position can be formulated by stating that some of these supposedly distorting mechanisms may constitute the agglutinating factor which keeps the group together and makes it efficient. 'Psychosocial scepticism' can also look at the functioning of society and adopt the form of the following query: 'Can a society with the type of mechanisms postulated for the "healthy" groups function? Will individuals find motives for acting if authoritarianism, rivalry, humiliations, elitism and aggressive ethnocentrism are eliminated? Or will the result be chaos and indolence, cultural inertia and social disintegration? Where would individuals find their motivation for confronting their work, the gruelling obligations of life? How would discipline be organized, etc?'

The reply we give to all these queries is related to our view of the human being, to what we believe about his profound needs and to our model of growth or psychic development. Psychological theories are sharply divided as regards these questions, showing a deep gap between different anthropological assumptions. It is my opinion that it was this gap which separated Freud from Reich, who was unable to accept the sceptical theses maintained by Freud in *Civilization and its Discontents*. My impression is that contemporary anthropological-psychological thinking is becoming polarized around two radical positions: one, a reductionist, sceptical instinctivism, and the other a humanist-existential approach.

Personally, I believe that healthy groups not only can exist but that they are an irreplaceable source of health and personal growth. And that to the extent that healthy groups are multiplied, the existence of humanity on the earth will change fundamentally, a new era will begin for humanity. Naturally, all this must be theoretically grounded and that is not exactly a small undertaking.

In addition, here the problem is posed of feasibility at the historical level of institutions which encourage or promote these healthy microstructures. The question obviously falls completely outside the limits of this chapter. But I also believe it constitutes a formidable, theoretical challenge for our times.

Notes

1 These groups may not be 'primary' in the original meaning of the term but they are at all events strongly significant for the subject. The category of primary group as opposed to secondary group alludes to a type of link with a high affective load. Cooley, the creator of this concept, characterizes it as follows:

> By primary groups I mean those which are characterized by intimate, face-to-face association and cooperation . . . From the psychological viewpoint, the result of this intimate association is a certain fusion of individualities into a common whole, so that the community life and the group's objective become the life and the objective of each one of them . . . Perhaps the simplest way of describing this totality is by saying that it consists of an *Us*: this implies the kind of mutual sympathy and identification which the *Us* expresses naturally. Every one of them lives with

the sentiment of the whole and finds it that sentiment the main objectives which his will sets for itself . . . .

The characterization which Cooley makes of primary groups indicates that the affective bonds involved by them are basically positive, empathetic and intimate. In principle this description fits the original family group (mother, father, brothers and sisters) – which in this paper we call the Primordial Primary Group (PPG) – since it is the first and most important for formation of the personality and also the family group which the adult can construct (spouse, children), a group of intimate friends, etc.

The 'primary group-secondary group' dichotomy is far from covering the whole spectrum of possibilities. Groups exist which are highly significant for the individual's life, where psychic interaction is strong and personal, but does not always respond to an empathetic or intimate nature. Thus, for example, for an adolescent his school group may not be a primary group as Cooley means it, because it may happen that an intimate link between him and his companions does not occur and there may be no collective identification with an empathetic Us, but it will always be significant and important in his psychic life. The adolescent will feel marginated or integrated, loved or rejected, successful or a failure, within a group. He will perceive among his companions a cordial or hostile atmosphere, competitive or co-operative, authoritarian or democratic and this, far from being neutral or indifferent for him, will have great significance in his emotional life, in his self-image, and in the perception of his possibilities of communication with the outside world. It is to this type of significance group, primary or otherwise, that we refer when we speak of the incidence of the small, non-primary groups in the development of the individual self. Our proposed terminology would be as follows:

Significant groups (Primary (Primordial / (Non-primordial
(Non-primary

2 References to these mechanisms made within the framework of different theoretical approaches can be found in numerous publications. We shall mention here only a few titles which seem to us particularly relevant for their detailed analysis of one or some of these mechanisms and their effects on human interaction or on individual personality:

Laing, R., Philipson, H. and Russell Lee, A., *Interpersonal Perception. A Theory and a Method of Research*, London: Tavistock Publications Ltd, 1966.

Berne, Eric, 'The Psychology of Human Relationships,' *Games People Play*, New York: Grove Press Inc., 1964.

Laing, R.D. and Esterson, A., *Sanity, Madness and the Family*, London: Tavistock Publications Ltd, 1964.

Watzlawic, P., Helmich Beavin, J. and Jackson, Don D., *Pragmatism of Human Communication. A study of interactional patterns, pathologies and paradoxes*, New York: W.W. Norton & Co. Inc., 1967.

Goffman, Erving, *The Presentation of Self in Everyday Life*, New York: Doubleday & Co. Inc., 1959.

Rogers, Carl, *Carl Rogers on Encounter Groups*, New York: Harper & Row Publishers, Inc., 1970. 'A Theory of Therapy, Personality and Interpersonal Relationships, as developed in the Client-Centred Framework' in Sigmund Koch, *Psychology: A Study of a Science*, Vol. III, New York: McGraw-Hill, 1959.

Kafka, Franz, *Letter to my Father*.

Janov, Arthur, *The Primal Scream*, New York: The Putnam Publishing Group, 1970.

3 I have developed this concept of maturity growth in 'Towards a Model of Human Growth', paper presented at the UNU-GPID Project meeting, Geneva, October 1978.

4 Regarding the so-called 'Encounter Group' movement, see Rogers, C. *Encounter Groups*, op. cit., and Schutz, William C., *Here comes everybody. Body, mind and encounter Culture*, 1971.

7
The Development-Adaptation Dialectic

Oscar Nudler

Introduction

The subject of this chapter is the dialectic between human development and social adaptation in modern society.[1]

In this first, introductory section, we shall consider our two key concepts in a general way. In the next section, which constitutes the core of this chapter, three forms of the development-adaptation relationship are introduced and illustrated. In the final section, a paradox posed by such relationship is briefly discussed.

We shall accept here that a need to develop or grow exists in every human being. We do not conceive this need only as a theoretical fiction or logical device but as a real impulse or force which motivates the human being to acquire new powers and capabilities and to perfect them. In the first years of life it is a dominant force whose existence is fairly evident to any observer of the child's rapid progress. Later on, once the basic accomplishments which define the human condition have been mastered, the impulse to develop is usually more difficult to detect, but this does not mean that it disappears. What happens is that it becomes part of the complex system which is the adult human being, a system through which it can continue to exert its influence openly or in which this influence can be counterbalanced up to a point at which its effect is annulled. Here we shall deal with the need to develop in those stages of life when it is no longer accompanied by the force of biological growth. Developmental psychologists do not in general concern themselves with these stages, assuming that development ends when organic growth stops. The reasons for this attitude are in part ideological – prevailing values are against continuous human development – and in part due to the intrinsic difficulty of the subject: the gains in excess of those which are basic to the species and the culture are more subtle, more difficult to observe and assess.

Our second assumption is that human development is a process intimately connected with another process: socialization or basic social adaptation. As a matter of fact, being on open system, every living creature requires an environment to exchange matter, energy and information. Without a minimum amount of adaptation to it, not only its development but just its mere survival would be impossible. The human being is certainly not an exception to this law of life. Due especially to the long period of helplessness of the infant, human development depends even more heavily on a successful adaptation. The essential *sine qua non* character of socialization for human development has become almost a truism, at least in the social sciences field, in contrast to our first assumption on the existence of an impulse towards continuous development. It is not so commonly accepted, however, that development and adaptation are not linearly associated to each other so that it is not necessarily true that to more (or less) adaptation corresponds more (or less) development, and vice versa. There exists, no doubt, a degree of social adaptation – basic socialization – which is necessary for development but, beyond it, an overemphasis on adaptation probably implies the non-actualization of important human potentials. It is clear, anyway, that adaptation should be sharply distinguished from development (or health or maturity or any other similar term).

In the process of development the human being constructs his own self at the same time as he builds his adaptive niche. Representations or images of the social environment, especially of 'significant others', become internal parts of the individual.[2] It is therefore somewhat misleading to phrase the problem of human adaptation in terms of the individual-environment distinction, as if two clearly separate realities were involved. But it is also wrong to assume, as happens frequently with the sociological approach to the individual, that the person is only the system of the parts he has internalized in the course of socialization. On the contrary, every human being is a *centre* of experience, an original form of feeling, of moving, of looking, of eating, of conversing ... He cannot be reduced to his biological heritage or sociocultural influence or to the sum of both. Even though biology and culture are the two pillars of the human building, every person shows some variable amount of irreducible originality.

The human being requires, as we said before, an environment, but he transforms this environment through his development into a *world*. There are important differences between the concepts of environment and world. The former requires spatial and temporal co-existence between its constituent elements, while the latter does not. A person's world is made up not only of perceptions of present events but also of memories of events in the past and of anticipations of the future. And of pure fantasized elements too. There may be a tremendous discontinuity between these elements. The human world deviates far from the actual environment with the help of mechanisms of symbolic transformation: natural and artificial languages, imagery, dreams, etc. All this is summarized when we say – recalling Ernest Cassirer – that man is a *symbolic* animal and that therefore his world is impressively different from the environment (or non-symbolic 'world') of other animals.[3] This surely makes biological[4] or zoological[5] reductionism untenable but does not constitute sufficient ground for denying the fundamental continuity between man and other animals based on the fact that he never escapes, in spite of his symbolic ability, from the basic facts of life: birth, growth, adaptation and death.

Having thus outlined the background of our approach, we can now proceed to ask how the development-adaptation relationship appears in modern society. We shall introduce a distinction between three main types or forms which this relationship might take, namely, predominance of adaptation, failure in adaptation and transcendence over adaptation. We shall describe these types in the next section. Let us just say here that each type comprises a varied, complex array of features of which we will take only a few, and that the human types which we will describe are pure, so that real human beings do not correspond exactly to any of them, although they might be closer to one type than to another.

Types of the development-adaptation relationship

Predominance of adaptation

The individual of this type is characterized by his high degree of adjustment to the average expectations of the members of the groups to which he belongs. He tends to play his parts efficiently and in each case he adopts the attitudes and conduct expected of him. Even his possible violations of social norms are 'normal' (and sometimes secretly approved and even admired). Adaptation does not often consist of an absolute adjustment to the explicit norm but rather of the apt handling of a mechanism composed by a formal facade and a hidden content. To take an example, the ruses to hide the extra-marital adventures of a husband are perfectly 'understandable', provoking knowing smirks and gossip. Countless plays have been written around this theme, which is still a lure to attract a large audience. The inverse situation, however, the case of the woman who deceives her husband, stirs up completely different reactions. It is rarely acceptable as a subject for light comedy: its natural place is drama. What happens is that in this case there is no duality, there is no implicit rule in the normative social code which makes tolerable or even encourages violation of the explicit rule. Nevertheless, and this is the point we wish to make here, the behaviour of the unfaithful wife, although it is not approved, is not seen as abnormal or strange. But there are other kinds of transgressions of social norms which are incompatible with a 'good' adaptation. Let us imagine, for example, the case of a couple which has adapted the so-called 'open marriage' rules and at the same time seem to be happy together and do not even dream of separation. Conduct of this nature would be regarded, particularly by non-intellectualized sectors, as utterly abnormal, suggesting some psychological or moral disorder. We can see therefore that along with the distinction between those transgressions which are legitimized by the underground moral code and those which are not, we have incorporated an additional distinction between 'normal' and abnormal violations. The hippie movement, before being extensively co-opted, provides a good example of collective abnormal violations of social norms and also of the strong negative reaction which this kind of behaviour triggers in over-adapted people. Generally speaking, the nature and seriousness of social sanctions against abnormal transgressions seems to be a good indicator of the degree of freedom allowed to its members by a given society.

Let us now approach more closely the individual in whom adaptation is the dominant force. His face, his posture, his smile, even his anger, are agreeably familiar. His picture reminds us of the stereotyped product of fiction created for mass consumption. However, a real human being, no matter how adapted he might be, could never attain the transparency or lack of thickness of commercial fiction characters. He always has something which does not quite fit into the prototype. What is it, how can this part which resists over-adaptation be characterized, the part that the over-adapted individual tries to hide and, if possible, suppress? According to Freud, that part corresponds to sexual impulses, especially when they are directed towards objects regarded socially as taboo. Freud's discovery of the deep split within human consciousness between an adapted and an unadapted part was remarkable, a discovery which contributed decisively to self-knowledge in modern society.[6] But he did not go up to the point of recognizing that the repressed part of a human being is *variable* according to different cultures and that it does not necessarily concern sexual impulses in the first place. For example, the greatest effort of social repression, at least in Western post-Victorian, mass consumption societies, seems to be directed more against what is genuinely individual than to what is common to the species, more against identity than against sexuality. What over-adaptation would repress in this case is, in short, the person's impulse to grow. There is no simple, clear-cut option, however, between over-adaptation and full development. Apparently minor or 'normal' decisions in one's life accumulate up to a point in which a sort of wall or barrier might stand in the way of any further real growth. Once their adaptive niche has been built, most people prefer to keep it, even though it is a painful prison, rather than running the risk of demolishing it to build a new one in its place. There are reasons which explain this apparently 'irrational' behaviour. An adaptive niche is a complex, delicate construction, which begins at birth or may be at conception. It is not a mere accumulation or a weak, loose articulation of parts. On the contrary, it is a structure endowed with multiple reciprocal relationships between the parts, so that to change some of them could have vast, deadly consequences for the rest. And growth, of course, implies change. Not only the great values and ideals are at stake in a process of real growth but also little habits, little stereotyped forms of conduct which reassure us that 'we have our feet on the ground'. And 'the presentation of the person in everyday life', the reflection in the mirror, the perception of people and things around us, in sum, the various elements which make up persons and their worlds. Therefore, once the adaptive niche has been consolidated, there is more often than not spontaneous resistance against change. However, the impulse to develop is deeply rooted in the human condition and when it is blocked it manages to make its voice heard. Nightmares, a varied array of physical disorders, irritations, anxieties, are some of the multiple ways in which the unhappiness of a being whose development is blocked, is expressed. To silence that voice, to refuse to hear it, the over-adapted individual resorts to alienation, in other words, to a behaviour designed to shore up obsessively the walls of his adaptive niche. His reasoning consists of telling himself that his dissatisfaction derives from not being totally consistent with his goals and he therefore proposes to be so. He hopes that by devoting himself completely to a given role – parent, businessman, sportsman, etc. – he will obtain greater success and recognition than he has had so far and he thinks this success will give him the happiness which eludes him. He locks out all other alternatives, all possibilities alien to his internalized model. Deep communication, real dialogue with such people is desperately difficult, if not impossible. Hence the particular impression of rigidity, of impenetrability, or as Marcuse would say, of 'one-dimensionality', which this type of individual produces in persons characterized by a more open psychological structure.

The type which we have been describing so far bears a more or less close resemblance to some well-known types which have been described by David Riesmann, Erwin Goffman and others. To conclude this section we will refer to the 'adaptation syndrome' as has been proposed by the American psychiatrist, Silvano Arieti. On referring to a patient who came for treatment for having areas of total anaesthesia on his skin, Arieti says:

> Obviously Mr Q could not reduce his whole life – his genuine existence and his potentially rich personality – to the level of adjustment ... In order to adjust to the level of adjustment (and this is not a play on words), he had to give up a great deal; he had to transform his

whole personality, so that in some aspects he became a caricature of what he could have been. The reshaping of his self required not only compulsive activity, but also a form of alienation, or depersonalization, so that he could not even experience the unpleasant aspects of this transformation. The hysterical anaesthesias were in a certain way a concrete representation of his depersonalization. The skin anaesthesias of which he was fully aware replaced another kind of anaesthesia which existed at a higher level and of which he was not at all aware: the repression of that part of living which aims at spontaneity and creativity.[7]

This case illustrates clearly the extremely negative impact which over-adaptation can have on human development.

Failure in adaptation

It is fairly obvious that modern society is not only constituted by adapted or overadapted individuals. A second, very important category corresponds to those who, even though they might have wished to adapt, have for some reason failed to achieve a 'satisfactory' level of social adaptation. This is such a vast, deep problem that it is even difficult to broach it. In principle we could divide the people to whom we wish to refer here into two large groups. First, the persons who, mainly because of the conditions in which they spent their infancy, have not achieved sufficient socialization in order to meet socially defined minimal standards of normality. Mental illness, criminal behaviour, or less spectacularly, a critically low degree of motivation, are often the indicators of such a situation. The second group of people showing adaptation 'failures' would be of those who, after achieving — unlike the others — an adaptive niche considered to be 'normal' within their culture, have suffered its destruction. We say 'have suffered' because this does not, as in the case of transcendence of adapatation which we shall discuss later, imply a destruction for the sake of growth, i.e. a new construction. On the contrary, it implies a destruction imposed by the external world, not desired — at least consciously — by the subject. The first group of persons has long been studied. On the one hand, physical factors have been identified which lead to failure in adaptation, especially those factors connected with extreme poverty, such as undernourishment in the first years of life which can irreversibly affect brain development, lack of medical care, hygiene, etc. On the other hand, psychological factors have been pointed out as responsible for subsequent failures in adaptation such as lack of early affection and stimulation, non-acceptance by adults of the child's personality, abandonment or overprotection, sex, class, ethnic and other forms of discrimination, family disintegration and so on. All this has been and continues to be widely studied so we shall not go further into the subject.[8] We shall only underline that our concept of failure in adaptation should not be confined to the analysis of individual cases but should lead us into the social system. We will recognize then that the failure in the adaptation of a human being is not independent of the adaptation of other human beings, just as honesty or chastity are not independent of prostitution, or poverty of richness. They are complementary aspects of the same system. For example, the family can be, as has been shown, a system structured in such a way that the 'normality' of the whole rests upon the 'abnormality' of one of its members. The weak, 'sick' member is a point of comparison which the others need in order to feel themselves strengthened in their 'normal' roles. The family thus becomes a trap for the discriminated member, a trap which does not allow him to escape from his assigned role and build up a different identity.[9] Other social micro-systems, such as asylums, prisons, hospitals, are also likely to contain within their walls 'trapped' beings, whose recovery is declared to be desired but who in fact are punished through marginality.[10]

The second group of persons to whom we want to refer here are, as anticipated above, those who have experienced a relatively tardy failure in their adaptation although, of course, their problem can usually be traced back in their personal history. In general, these are people who have achieved an early, closed adaptation and have clung to it, refusing any subsequent change. A typical example is the woman who dedicates her whole life to the care of her children and once they become independent – if they manage to do so – she starts to feel an acute lack of existential meaning. The underlying illusion which finally fails to become true is that of growing through the growth of her children. But the task of personal growth cannot be transferred and the attempt to do so might be harmful not only to the 'donor' but also to the supposed recipient of the 'donation'.

Another very frequent case of late failure in adaptation in modern society is supplied by those who, after being forced to retire from the working world, do not know what to do with their free time and, more generally, with their lives. There is a fundamental analogy between this case, the previous one and many other examples of late failure in adaptation. In all cases, the individual has lived a sufficiently long period of life in a state of closed adaptation. All of a sudden some circumstances beyond the individual's control – decisions of other people or the impersonal action of social rules – have demolished the adaptive niche and have left the individual out in the cold, so to speak, at a stage of life in which he or she feels it exceedingly difficult to start again.

There is a basic similarity between the two group of persons who fail in their adaptation: those who fail early, who have never achieved sufficient adaptation to be considered normal, and those who fail late, who have lost the adaptation attained without achieving another. Both share the same problem which consists of a serious deficiency in their self-image. A question commonly heard in states of crisis from people of this type is: Who am I? This question is not motivated by a philosophical purpose of self-knowledge but instead it expresses a feeling of lack or loss of a clear insertion in the world.

We said before that the process of construction of a person's world is inseparable from the construction of the inner self. This explains why failure in adaptation, that is, the lack of an adequate adaptive niche, is linked to identity problems. However, a deep questioning of one's identity does not necessarily imply failure in adaptation. On the contrary, it might indicate a transition towards a superior form of adaptation to the world. Let us recall, for example, the way in which Tolstoy described himself in his fifties: 'I felt that something had broken within me on which my life had always rested, that I had nothing left to hold on, and that morally my life had stopped . . . I did not know what I wanted . . . why should I live? . . . Is there in life any purpose which the inevitable death which awaits me does not undo and destroy?'[11]

Many examples might be cited of persons who, whatever the criterion we use, would probably be considered as cases of good human development and who, at the same time, have undergone in their lives deep crises of the type just described. These are persons who can overcome and even use such crises for their growth. We will refer to them in the next section.

The illustrations of failure in adaptation we have given were chosen for their typical nature. It must be remembered, however, that failure in adaptation does not always have such clearly perceivable causes as in the examples chosen. Sometimes it is due to obscure factors which persistently elude any simple explanation. There are cases, for example, in which apparently there is nothing in the individual's life which would sufficiently explain his sinking into a state of paralyzing, even regressive crisis. Percy Knauth, an American journalist, has left us an account of his thoughts and feelings when he went through a period of profound depression. After a long time in total darkness about the causes of his state, Knauth advanced some guesses but it is apparent that the crisis remained for him as a basically obscure, incomprehensible episode.

> What had happened to get me to the point when I only wanted to die? It was as though some silent dark force had pulled the plug out of my life and let the vital juices drain away until I was an empty shell.[12]

There are many Percy Knauths in modern cities but there are many more people who do not go through such extreme crises but who, like him, feel a deep dissatisfaction whose main roots remain elusive for them. It is certainly difficult to recognize under such unhappiness one's own failure to actualize that unique possibility which is every human being.

Transcendence over adaptation

The human type we intend to consider briefly now overlaps to a great extent with various models of the healthy personality which have been proposed by among others, C. G. Jung (the individuated person), E. Fromm (the productive character), C. Rogers and A. Maslow (the self-actualizing individual), etc. As was stressed above, social adaptation is not a sufficient condition for human development and, as happens in the case of over-adaptation, can even go against it. Personal growth (or human potential) theorists have insisted on the insufficiency of adaptation: 'Supporters of the human potential movement suggest that there is a desirable level of growth that goes beyond "normality", and they argue that it is necessary for human beings to strive for that advanced level of growth in order to realize, or *actualize*, all of their potential. In other words, it isn't enough to be free of emotional illness; the absence of neurotic or psychotic behaviour is not sufficient to qualify one as a healthy personality. The absence of emotional illness is no more than a necessary first step to growth and fulfilment. The individual must reach further.'[13] However, transcendence of adaptation does not imply, at least in the sense we understand it here, cultural marginality or isolation. The form of transcendence in which we are interested is that of those individuals who have managed to grow as persons *inside* a culture or sub-culture which, although it provides some opportunities for human transformation, raises at the same time huge obstacles or tempts with non-transformative sidetracks. The transcendent person as understood here thus has deep roots in the culture but, unlike the types described previously, he is not a mere reflection, a passive plaything of social dynamics. He is able to become aware of and to resist influences coming from the very culture in which he has grown. This ability to resist externally induced 'outside-in' change is not opposite but complementary to the capacity for inner 'inside-out' change. Both abilities are certainly not acquired in one day, but they are the outcome of a long liberation process through which the individual 'personalizes' his cultural heritage and makes it part of a unique, integrated personal whole. Individuals belonging to the previous types are, on the contrary, somehow internally dissociated, either because the adapted part has grown excessively, stifling other potentialities, or because a sufficiently strong central core capable of organizing the personal structure around it does not exist.

Integration does not exclude conflict; it means only the possibility of a deep dialogue with oneself. Arthur Janov, in characterizing the mature person, says that he is one 'who thinks what he feels and feels what he thinks'.[14] We would not go so far as that since we believe that inner tension between ways of thinking and feelings is part of a mature or healthy personality. However, this conflict should neither result in a denial of one's own feelings nor should it reach the point of becoming destructive. In other words, a tendency towards awareness and unity – through differentiation and integration – of the inner self is dominant in a healthy personality.

Following Maslow, in referring to transcendence we do not have in mind geniuses or even admittedly outstanding individuals. Social recognition does not usually have a high correlation with belonging to this category. They could be not very 'successful' individuals, they might not shine in any conventional sense. Neither is it necessary that they should be individuals equipped with exceptional psychological powers. Our type is not that of the individual who is always smiling whatever the circumstances. A mature person who lives in a social environment full of inequities and obstacles to human development cannot avoid being affected by them since openness and sensitivity are precisely among the traits which contribute to his maturity. However, he would not take refuge in a private world but would try to contribute to the solution of the concrete problems which crop up around him. Gandhi's term *Swadeshi* expresses very well this idea of loving concern about what surrounds us.[15]

An additional feature of the type we are describing which is worth mentioning is his attitude towards uncertainty. Instead of denying uncertainty, or accepting it with resignation while trying to accommodate to it, this type is characterized by an active search for new development patterns with its associated increase of uncertainty. Thus, the search for security is not, as in the previous types, the key determining factor of attitudes and behaviours.[16]

Finally, a transcendent person is not only inclined to trust others but also to accomplish his development together with them. In other words, his development is at the same time *co-development*.[17] In a previous study we have distinguished three fundamental needs, namely, the needs for identity, development and transcendence.[18] We claimed there that these three needs make up a system so that none of them can be properly met if the others remain unsatisfied. We have focused in this chapter on the need for development but we had to refer to the other two as well. So we have seen that the lack of a sufficiently strong identity or the exaggerated linkage of one's identity to social adaptation constitute decisive restraints to development. Now we can add that transcendence towards the world, particularly towards the human and social world, is also a fundamental requirement for human development.

Concluding remarks

We shall end this chapter by making two remarks and one question. The first remark is just a consequence of what has been said before about the fact that real persons should not be equated to pure types. Although transcendence over adaptation represents our desirable development type, this does not imply any belief in the existence or need of an elite of 'transcendent' individuals qualitatively differentiated from the rest. We only assume a relative predominance in each person of one of the typological clusters we have described. Even for the same person, the balance can change radically throughout his life cycle so that, for example, he can pass from the dominance of overadaptation or marginality to greater transcendence or vice versa.

Now we want to insist – this being our second remark – upon the fact that all the problematique discussed in this chapter makes sense only in a culture in which a sharp distinction between individual and society takes place. In cultures where this is not the case, the study and assessment of human development might require, of course, a totally different analytical framework.

Lastly, let us turn to a paradoxical question referred to the individual-society relationship: is a society which fosters systematically continual development of its members logically possible? The paradox which is implied here is as follows: if it is true that full human development requires some amount of non-adaptation to society or, more exactly, transcendence over adaptation, then a social order which fostered systematically the continual development of its members would promote by the same token its own disappearance. At least in the long run, such social order would not be self-sustainable.

A classic solution of this puzzle has been to reduce as much as possible the number and scope of social norms by making them entirely flexible and open to change. But this solution presents, in the first place, the difficulty of reaching a consensus on a definition of a minimal threshold beyond which we would have just *anomie* or even dissolution of the social order as such. A second difficulty, even more crucial, is that the human being needs a structured social environment for attaining basic socialization and also for struggling towards transcending it. A normless, distructured, anomic society would probably be at least as unfavourable to human development as a society with an excessive number of norms.

An alternative solution, which has been advocated, for example by J. Galtung, follows from a vision of an ideal society in which opposite life-styles would co-exist and would be open to free individual choice.[19] In such a scenario, the individual–society clash is easily avoided since the individual who feels unsatisfied with a given life-style can change it for another one. He could, for instance, move from an individualist social context in which privacy is the higher value, to a community setting in which sharing is the guiding principle, or vice versa.

However appealing, the conceptual difficulties of this proposal are at least as great as in the previous case. On the one side, the required co-existence of separate and opposite life-styles would imply the co-existence of separate and opposite economies, education systems, value systems and so on, something which is difficult to accept as a workable possibility. It is true that to some extent this is what actually happens in modern society where, for instance, there is a non-formal economy along with a formal one. But these different sectors are linked in such a way that they should be considered as complementary parts of the same system, no matter how conflictive their relationships might be.

On the other side, even if we grant the theoretical feasibility of the extremely pluralistic model we are discussing, its dubious advantage would most likely be the substitution of the individual–society clash by a conflict between different social sectors. And if one wanted to prevent this conflict by excluding ethnocentric, non-tolerant groups, one would deny the pluralistic starting point and would foster the reappearance of the individual–society clash.

The unsatisfactory nature of this and other proposals is maybe rooted in a wrong epistemological approach to the paradox posed by the development-adaptation dialectics. The epistemological mistake involved would lie not so much in the proposed solutions as in the very attempt to eliminate the paradox so as to get a perfectly harmonious individual-society relationship. A better approach should accept the inescapable paradoxical nature of this relationship and should then try to explore and use its mechanisms and consequences.[20] It is true that in its present form such a relationship is more harmful than favourable to human development and, therefore, every effort to modify it is more than justified. But this does not mean that in a more desirable social order all contradictions between individual development and social order would disappear. It only means that such contradictions would appear in new, more creative forms, capable of impelling human development in directions which we, from the lookout of modern society, can just poorly glimpse.

Notes

1 Quite different aspects might be picked up to define the classic concept of 'modern society'. Given our concern here, the more relevant of those aspects would perhaps be the dominant role of scientific and technological rationality, closely associated to a strong emphasis on economic growth and industrial development, based on an extended use of non-renewable sources of energy and natural resources; the disruption or even total destruction of traditional community links, the emergence of bureaucratic, 'rational' organizations, private or State-owned, characterized by a built-in tendency to limitless growth; and the constitution, particularly through mass media, of an homogenizing, all pervasive symbolic milieu.
2 Mead, G.H., *Mind, Self and Society*, Chicago: University of Chicago Press, 1934.

3 Cassirer, E., *Essay on Man*, New York: 1944.
4 See E. O. Wilson, *Sociobiology: The New Synthesis*, Cambridge, Mass.: Harvard University Press, 1975 and the critique of sociobiology made in Harris, M., *Cultural Materialism, The Struggle for a Science of Culture*, New York: Vintage Books, 1980.
5 A well-known example of this form of reductionism is *The Naked Ape*, the best-seller written by Desmond Morris.
6 Actually, almost all scientific discoveries have a long history behind them. In the case of Freud, see White, L.L., *The Unconscious Before Freud*, New York: Basic Books, 1960.
7 Arieti, S., *The Intrapsychic Self. Feeling and Cognition in Health and Mental Illness*, New York: Basic Books, 1976, p. 240.
8 See for example in this connection the classic studies on 'hospitalism' by J. Bowlby, *Child Care and the Growth of Love*, London, Pelican Press, 1953 and R. Spitz, *The First Year of Life*, New York: International University Press, 1965.
9 See Laing, R.D., *The Divided Self*, London: Tansbook, 1960.
10 See Goffmann, E., *Asylums. Essays on the Social Situation of Mental Patients and Other Inmates*, New York: Doubleday, 1961.
11 Tolstoy, L., *Confession* (1882), quoted in Schultz, D. *Growth Psychology: Model of the Healthy Personality*, New York: Van Nostrand, 1977.
12 Knauth, P., *A Season in Hell*, New York: Harper & Row, 1975.
13 Schultz, D., *op. cit.*, p. 2.
14 Janov, A. and Holden, M., *The Primal Man: The New Consciousness*, New York: Thomas I. Cromwell, 1975.
15 Gandhi says: 'After much thinking I have arrived at a definition of Swadeshi that perhaps best illustrates my meaning. Swadeshi is that spirit in us which restricts us to the use and service of our immediate surroundings to the exclusion of the more remote. In the domain of politics, I should make use of the indigenous institutions and serve them by curing them of their proved defects. In that of economics. I should use only things that are produced by my immediate neighbours and serve those industries by making them efficient and complete where they might be found wanting' (M.K. Gandhi, *Speeches and Writings*, Madras, 1977).
16 Gallopin, G. 'Personal Attitudes Towards Uncertainty and Surprise', Fundación Bariloche, Bariloche, 1981.
17 For a more detailed account of the concept of *co-development* see the chapter by C. Mamali and G. Paun in this book.
18 Nudler, O., 'Human Needs: A Sophisticated Holistic Approach' in Lederer K. (ed.) *Human Needs: A Contribution to the Current Debate*, Cambridge: Oelgeschlager, Gunn & Hain, 1980.
19 Galtung, J., *The True Worlds*, New York: The Free Press, 1980. ch. 3.
20 Professor S. Marcus, from the University of Bucharest, who has made a comprehensive study of paradoxes in different domains, has convinced me of the fruitfulness of this point of view.

Section IV
Human Development: Regional Perspectives

8
Human Development in Contemporary Industrial Societies

Ian Miles

Introduction

To begin with, some data from polls carried out in one industrial country, Britain, in 1979: four out of five Britons thought the country was a reasonably good place to live in, but more than half of them thought things had got worse over the 1970s, and over a third thought that the next decade would see more overall change for the worse. Despite two-thirds of the sample feeling that the past decade had brought them more free time, and over 80% that individual freedom had increased, the majority feeling, that things will change faster in the future than now, was not matched by an optimistic view of change. Large majorities felt violence and lawlessness, noise and pollution, crime with money and material possessions, and an unwillingness to work hard to be on the increase; more people thought honesty to be decreasing, government bureaucracy and restrictions to be increasing, than did the reverse. People with low incomes felt that their circumstances had declined; young adults joined them in pessimism concerning the future.[1]

To pile on the sense of social decline, focusing especially on interpersonal behaviour, a poll carried out in early 1980 may be cited. Here, a large majority (over 70%) of Britons felt people had become more aggressive and selfish, less polite, honest and moral; while open-mindedness and knowledgeability are believed to have increased by 82% and 69% respectively, tolerance and kindness are believed to have declined by the majority. Attitudes to many issues of public morality (for example, whether too much violence is portrayed on television, which on the whole is felt to be the case) diverge quite markedly across age, gender and educational level. But it is evident that a large majority would claim that our society is a less honest, secure, compassionate place than in the recent past.[2]

Yet this is a country whose population was identified as amongst the most happy in the world, by a Gallup survey involving almost 70 nations. Ninety-two per cent of Britons in 1976 claimed to be either 'very happy' or 'fairly happy' – behind Scandinavia (here treated as one country) and Australia, and practically tying with Canada. This cross-national study confirmed earlier work tending to show that self-reported happiness and satisfaction are higher in richer (and thus more industrial) countries. It should be noted, however, that while contentment with living standards and most aspects of ways of life is cited by people in richer countries, questions dealing more with wishes than with attitudes tell a rather different story: thus one in five Britons said that they would like to emigrate, four-fifths would like to leave big city life for more rural surroundings, and three-quarters reported no desire for continued industrialization. (In most poor countries, industrialization was seen to be the key to a better future; and in almost every country the majority felt that their lives had improved over the past five years, and would continue to do so).[3]

There are numerous ways to account for such results. Let us first dismiss the idea that we are leading with the 'peculiarities of the British', that the marked nature of economic decline in Britain means that Britain is an unusually mixed-up country in so far as social attitudes are concerned.[4] We do not wish to minimize the significance of natural differences within the very heterogenous set of 'industrial countries', but certain features seem to be quite common (at least in OECD countries, where polls are more widely available). The study of *Images of the World in the Year 2000* demonstrated pessimistic attitudes to the future in a number of industrial countries, East and West alike.[5] An EEC-wide survey found majorities in France and Italy arguing that the future is too uncertain to take on parenthood (although these two countries had the greatest proportions agreeing that parenthood is a fulfilling experience).[6] In 1978, despite a considerable increase from the gloomy early 1970s when a minority of Americans expressed a high level of satisfaction with life in the US, a majority of Americans believed that their country would soon be involved in a war and that the government was inattentive to their needs, and a majority of 13–18 year-olds thought that the world would be a worse place to live in in the late 1980s.[7] Fear of crime has penetrated American life, with a majority of households owning guns, and two-fifth of the population reporting feeling unsafe in their everyday environments and anticipating criminal assault.[8] Fears of nuclear war have risen in prominence recently, too, as polls in many countries demonstrate.[9]

These examples, which could be multiplied manyfold, serve to indicate that we are not just dealing with a British phenomenon. And while concerns about the state of the economy, international relations and the like are clearly volatile, reports of pessimism concerning the future and a sense of deteriorating social relationships seem quite widespread over industrial countries.

Of course this could still mean several things. Perhaps people are parrotting what they are told by the mass media, and there is a systematic tendency there for a negative evaluation of present trends. Or perhaps people everywhere are likely to hark back to the 'good old days'.[10] These are the types of explanation beloved of those defending present directions of social development in industrial societies. On the other hand, social critics of conservative and radical persuasions alike are liable to take attitudinal data of this sort very much at its face value, taking it as direct evidence of objective deterioration of the quality of life and potential for human development in rich countries, whether this is to be explained by individual shortcomings (e.g. 'materialistic consciousness') or structural factors (e.g. the changed role of the state, or the extension of commodity relationships into new areas).

Whatever the case, it may be pointed out that such results can hardly bode well for psychological growth. If these surveys simply indicate media manipulation or nostalgia, the degree of contact of the mass of people with social reality is thrown very much into question. If they are fairly accurate representations of social trends then there is certainly cause for concern – unless the growing pessimism and distaste only reflect a developing sophistication and realism about social affairs. There is surely some weight to this last argument, however distasteful it is, if only because numerous studies of policy-making indicate that growing public cynicism concerning the ways in which the political system operates is well-justified.[11]

A growing sense of distrust, suspicion and fear in relationships between people has serious implications for human development. Even if these feelings may be partly warranted as defensive reactions, they are often likely to affect the behaviour of those confirmed in such a way as to render them self-fulfilling prophecies. The effort involved in breaking down barriers of suspicion is considerable: the response is likely to be one of retreat into formal communication; or reliance upon symbolic manifestations of life-style and attitudes to locate like-minded and/or trustworthy people; or pervasive hostility and attenuation in relationships. Why is this counter-productive for human development? A first answer to this is simply that the tendencies cited will ensure that a more limited amount of feedback on oneself, and one's impact on others, is obtained. A vital source of information on one's actions and the internal life of others is diminished: to the extent that psychological growth depends on such information (and we would argue that the relationship is intimate), opportunities for growth are restricted.

Furthermore, it is not unreasonable to argue that human development is itself related to feelings of trust, fear of aggression, and the like.

An important feature distinguishing human beings from most animals is our capacity to empathize with and understand others' lives; our development of social needs that extend beyond the self-interest of individual identities. Given this, it may well still be the case that at any point in time an individual's maturity will be accompanied by a realistic distrust of some fellow beings, a cynicism about the benevolence of major social institutions, and the like. But over time, an improvement in the level of human development would be expected to be both reflected in, and facilitated by, increased feelings of trust and concern for others' welfare – although the process might often be a far from smooth one!

To round off this introduction, let us briefly refer to some other types of empirical data. A related project to the GPID project of the United Nations University has been concerned with charting costs and benefits of dominant ways of life, and possibilities and practices of alternative ways of life, in industrial societies.[12] Two conclusions are relevant to our immediate concerns.[13] The first is that in terms of many indicators of aspects of psychic well-being – psychiatric illness,[14] suicides, addictions, violent assaults (including child abuse), stress-related deseases – the situation in industrial countries is generally deteriorating. Clear trends are evident in this data, at least, although its interpretation is often open to doubt – it may be that it is easier to commit suicide or receive psychiatric treatment now; medical and police reporting practices may have changed. The consistency of the results suggests that something more fundamental than the undoubted changing functions of the state with respect to psychic life is involved, however. The second conclusion is that there has been a growth in attempts to develop alternative ways of life, which reflects in part changed opportunities and in part a deepened sense of the limitations of the existing social organization of the industrial world as far as human growth is concerned.

These tendencies reflect more than frustration and fatalism born of the economic crisis which has been evident since the early 1970s. The recession may have reversed some trends and accelerated others, but it is still possible to look at the last few decades as being of a piece. Growing social malaise, as indicated by actions destructive of oneself or others, is accompanied by gloom about the future, and by trouble surrounding human relationships in the present social order.

However, the implications for human development are frequently contradictory. In the following sectors of this chapters we seek to assemble empirical evidence relevant to this issue, and more often than not find it ambiguous.[15] As with most aspects of the modern world, we can characterize the processes involved as resulting in *uneven development*: gains and losses of different kinds are accumulated in a constantly shifting pattern (although their distribution within societies typically shows a consistent unevenness in favour of more powerful classes and groups); the erosion of some traditional moulds of psychological development may produce casualties and opportunities for consciously-organized action at the same time.[16] Perhaps the main conclusions of this chapter can already be summarized: the picture of human development in industrial societies is complex, far more complex than is admitted by optimistic prophets of post-industrial society or by many radical critics of the oppressive nature of our social formations. We accept that there are many obstacles to human development located at the core of modern industrial societies, and that the structural change required to overcome these is a matter of, and for, conflict. But individual psychology does not passively respond to external circumstances: it is itself an area and source of struggle between opposing tendencies, so that potential for progressive change is at least latent within most of our conditions of life.

Assessing human development in industrial societies

The introduction has, hopefully, demonstrated that it is possible to make some statements pertaining to contemporary 'industrial societies' in general. However, levels of economic development, cultural patterns, and political organization are very diverse in different industrial countries. We shall not be able to spend much time discussing cross-national differences, although, for example, the monopolization of power by the bureaucratic state apparatus in the CMEA countries implies significantly different prospects for the exercise of some human abilities in these countries as compared to the OECD region in which centres of power possess rather more autonomy. There are many shared features in the technological and cultural organization of different industrial societies in the contemporary world, and this to a large extent derives from a particular historical formation of the world system. In dealing with these common features, then, we are not seeking to advance a theory of 'industrial society' *per se*: existing industrial societies are themselves very different, and the immediate future may witness divergence, especially if 'newly industrializing countries' maintain their momentum along existing trajectories. But much more vastly different industrial societies might well exist: 'industrialism' itself is shaped by a matrix of social relationships.

What then, shall we understand by human development, the other component of our brief? Nudler and Mallmann formulate it in terms of 'psychological development through an enrichment and expansion of experience' which includes in addition to material satisfaction and equity, 'opportunities for human encounters . . . for communication in small groups and communities and for the development of feeling as well as thinking' (p. 3).[17] The emphasis on feeling as well as thinking is important, but we would prefer to go beyond 'experience' itself: how is it to be enriched and expanded?

Perhaps the best approach to human development here involves thinking about what is specifically *human* in it. We would propose that a central feature here must be *agency*, and, in particular the awareness of oneself as a goal-directed being, as an agent.[18] What is agency?

The notion of agency as 'conscious, goal-directed activity' has been elaborated by Anderson (1980), who distinguishes between three types of goal.[19] First are private goals, individual choices which have preoccupied people throughout most of history, which reflect and typically reproduce existing social relations.[20] Second, and less common, are public goals which are aimed at achieving collective action within a known structural framework. Third, even less frequent and, with only isolated exceptions, a product of the modern era, are those collective goals embodied in projects whose agents seek control over 'their collective mode of existence as a whole, in a conscious programme aimed at creating or remodelling whole social structures' (p. 20).

Agency is not merely a matter of exercising control: it is more a case of taking charge of one's action. Human beings are specifically able to be aware of themselves as conscious, goal-directed beings, relating to other human agents. As we act in the world, we can understand our actions and their consequences: not merely rational knowledge of the world 'out there', but also apprehension and comprehension of our own feelings. Human development can be seen as the growth of agency: in people's individually and collectively consciously acting to create the circumstances of their lives. This implies solidarity and sharing of experience, and respect for others living now and in the future. It means the opportunity to participate in situations allowing one to experience: one's own capacities in the physical and social world; the evolving of emotional commitments of self and others, and how these affect the choice and confrontation of goals; and thus the nature of human goals and the sorts of action that can further human development.[21]

The task of assessing human development in terms of this definition is by no means easy. As Anderson notes, the growth of technological capacities and modern forms of class society has increased our potential understanding of the physical world and potential for collective acts of agency so as to change those structural circumstances of one's life that are not amenable to the agency of isolated individuals. Studies cited previously, however, suggest that such potentials are not presently expressed in greater subjective awareness of human possibilities. It is not possible here to attempt to investigate this paradox in detail, although we believe that it must be related to the failure or degeneration of revolutionary movements in industrial countries in the twentieth century and the consequent contortions of late capitalist and state-socialist formations. We will for the present seek to use a more empirical approach, at the risk of some narrowness and loss of theoretical power.

We can approach the problems of human development through the prisms of space and time. On the one hand we could take the major structures and institutions which create life-opportunities in industrial societies, and ask such questions as: What are the cognitive and affective relationships of individuals to these social forms? What is the degree of agency that they can exercise over them? On the other hand we might consider the daily life of individuals as a phenomenon structured in time, and ask: What are the areas of activity in which people are occupied? What opportunities do these provide for the exercise and growth of human capacities?

Time use constitutes a convenient starting point. The categories provided by time-budget researchers already direct attention to some social institutions, through defining such activities as 'work' and 'housework'. Other institutions will become apparent through their absence as clearly differentiated demands on time for most people, most of the time, although they actually structure a great deal of people's activity – for example, political institutions. The data to be discussed will largely feature adults – the experience of children is quite different (for instance, schooling is an important activity, and thus involvement with state agencies is actually more quantitatively visible) – but a focus on adults seems reasonable. Even though children have quite specific problems, any maldevelopment in the adult population should give a viable provisional view of the state of human development in industrial societies.

How is time spent in industrial societies? Time-budget research gives many insights, and the collection of studies edited by Szalai (1972) is particularly helpful[22] – even if the categories provided by earlier researchers are often quite clumsy for present purposes. (For example, sexual activities, and leisurely interpersonal relationships in general, are very much underplayed.) The Szalai studies, sponsored by UNESCO, cover adults in some eleven industrial countries in the OECD and CMEA in the mid-1960s, and it is interesting to note a considerable degree of similarity across them. Table 8.1 indicates, for example, that some 40–46% of the day is spent, on the average, in sleeping, eating and personal hygiene as the main activity, with a similar proportion of time shared more or less equally between paid work and leisure. If we take paid and unpaid work together (the later including housework, etc.), then this typically occupies from 31% to 44% of the day, with leisure activities taking up from 15% to 22%.

'Personal needs'

Clearly the single largest proportion of time is spent on 'personal needs', among which sleeping is the most time-consuming activity.

Similar results here would be expected for children, too, of course. For a number of reasons, the significance of this activity in human development is rather overlooked. It has received only specialized research, given that it is not an area of life in which people are obviously producing very much of commercial or political value, nor one in which they consume much more than beds and nightwear. The apparently irrational and sometimes disturbing content of dreams leads to their being repressed on a massive scale, for dream research has established the frequent nightly incidence of dreams in most people, while anyone who has experimented with dream recall can testify to the vivid mental activity that accompanies much of our sleep.[23] Perhaps the direct and indirect association of sleep with intimacy and sexuality has discouraged attention to sleep itself, too.

To assess the implications for human development of this domain of activity requires more detailed assessment than can be provided here. It may be useful to release some of the relevant arguments however. First, we would argue that the regular forgetting of our dream life is a severe loss. Not only is one block of time lost in memory, but also a whole range of mental activities is excluded from consciousness and thus from explicitly forming part of the self-image of the dreamer. This is an impoverishment which presumably contributes to the exclusion of real, but conventionally unacceptable aspects of ourselves from our self-awareness, thus concealing some root of our behaviour and rendering us less capable of resolving internal contradictions.

Second, however, we cannot simply assume that recall of dreams, or effort expended in their interpretation, necessarily moves us toward greater self-awareness. Who, for example, has applied more effort to dream recall and analysis in industrial societies than the psychoanalytic movement? Yet as Timpanaro points out in discussing Freud's account of slips of the tongue, there is no mention here of slips deriving from anxieties concerning either class relations or biological frailties.[24] While some notable post-Freudians have been active in liberation struggles, the great dreamer Jung flirted with Nazism, and Jacoby has scathingly outlined the course of conformist psychoanalysis.[25] Greater acceptance of, and ability to communicate, one's dreams – as with other repressed experiences and feelings – may be a necessary condition for further growth, but the way in which this material is used, the relations which it enters into and helps to reproduce or transform, are crucial.[26]

A third, and related point, concerns the extent to which we are here discussing a facet specific to industrial societies. Some pre-industrial societies seem to have devoted quite extensive effort to integrating dream experience into everyday life.[27] From a largely anecdotal knowledge of anthropological and historical evidence, however, the present author suspects that this is the exception rather than the rule, at least in the antecedents of contemporary capitalist and state-socialist formations. Further, while mythic symbolism may provide an analogue for

Table 8.1 Time budget for industrial societies

Percentage time spent:	Belgium	Kazanlik, Bulgaria	Olomouc, Czechoslovakia	Six cities, France	100 electoral Districts, Fed. Rep. Germany	Osnabruck, Fed. Rep. Germany	Hoyerswerda, German Dem. Rep.	Gyor, Hungary	Torun, Poland	44 cities, USA	Jackson, USA	Pskov, USSR	Kragujevac, Yugoslavia	Maribor, Yugoslavia
(1) Paid work (including travel to job)	20	28	23	19	17	16	22	26	23	18	18	26	19	22
(2) Housework and household care	12	10	15	14	16	15	17	15	13	13	13	12	14	19
(3) Child care	1	1	2	3	2	2	3	2	2	2	2	2	2	2
(4) 'Personal needs' (sleeping, eating etc.)	45	43	43	46	46	46	42	42	41	43	43	40	42	41
(5) Non-work travel	2	3	2	2	1	2	2	2	3	3	4	4	3	3
(6) Study	1	1	1	1	0	1	1	1	1	1	1	2	1	1
(7) Religion and organizations	1	0	0	0	0	0	1	0	1	1	1	1	0	0
(8) Mass media	9	5	8	6	6	8	8	6	8	9	9	8	6	6
(9) Leisure	21	16	17	17	18	21	16	14	18	21	22	17	22	15
Proportion of women employed (%)	41	84	70	48	36	44	70	60	66	49	48	92	43	61

Note: All percentages are rounded up. Differences among small figures therefore must be treated with caution. There also seems to be a systematic error in the original table, leading to a longer than 24 hour day – or in this case, more than 100% of time being used.

Source: Calculated from Table 1 of Robinson, Converse and Szalai (1972), in Szalai (1972) op. cit.

understanding psychic processes and social forms, some well-known dream interpretations in terms of supernatural messages, are rather suspect as a means of advancing psychological understanding. We might tentatively introduce Marcuse's concept of surplus-repression here, used to indicate that much repression of our psychic impulses relates not strictly to the needs of *any* civilized life, but specifically to the control of feelings and action in order to maintain existing structures of social domination.[28] With technological advance, he argues, less time has to be spent in necessary labour, fewer needs are conditioned by scarcity, and therefore repressive systems will either relax or engender increasing surplus repression. Freudianism could be interpreted as, in part, an attempt to relax the repressive controls around dreaming (just as it was also related to the 'revolution' in sexual customs in industrial societies), but its limited influence as validating dream experience as worth serious consideration – even into intellectual culture – would suggest that powerful self-reinforcing practices sustain surplus repression. The functions of sleep are not well understood, although evidence on dream deprivation suggests that dreaming is as important as physical rest.

The total time spent in activities connected with 'personal needs' is, on average, fairly similar across the countries sampled (differences between averages do amount to somewhat less than 1-1/2 hours). As for trends in these activities, the picture is rather unclear; Gershuny and Thomas (1980)[29] report small increases in the time spent in these activities in the US (1965–75), but a practically unchanged allocation in the UK (1961–75).[29] It would seem that male and female 'personal care' is diverging in the US and converging in the UK (although women still devote more time to this area).

It would be interesting to investigate the subcategories of personal care further. Time spent in washing, health care, eating, and so on could reflect obsessive tendencies, proper concern with self-managing important aspects of one's physical condition, or a desire to appear well in terms of media concepts of fashion and beauty. The emergence of the women's movement and alternative ways of life movements, respectively, have brought these 'personal' issues into the domain of political debate in the West – or rather shown that the choices made here can reflect and reproduce social relations of inequitable power. While the slogan 'the personal is political' may divert attention from the specificity of the 'personal', and in any case has been most applied to interpersonal relationships, its salience here is demonstrated by interventions into the question of diet by feminists, ecologists and radicals.[30] Some of these groups have undoubtedly had some success in raising consciousness about issues of diet, health and the like (as manifest by the extent of sales of books on these topics, on the growth of a 'whole food' industry and of alternative self-help medicine, yoga, etc.), but the extent of their influence, and the degree to which the terms of the debate are being changed, rather than a rearguard action being fought against commercial and destructive tendencies, is less easy to assess.[31]

Paid and unpaid work

Work is a central feature of most people's lives in industrial societies. Of the population aged 15–64, an average of 68% were 'economically active' in the six most dynamic OECD countries, 64% in eleven other OECD countries, and 77% in seven CMEA countries, in the mid-1970s.[32] Female employment is usually higher in the state-socialist countries (see Table 8.1), but has been increasing over the postwar period in most Western countries. Almost all Swiss married women work, while in several other countries the figure is well in excess of 50%; but in a few countries, of all women of working age, less than a quarter are in paid work.[33]

Working hours also vary considerably. Workers in manufacturing for the three groups of countries cited above, work on the average 40.6, 38.8 and 42.1 hours per week. It would seem that the amount of time spent in paid work has typically declined in industrial countries.[34] Of course, the data cited is from periods before the current recession, which has thrown many workers in the West into involuntary unemployment with associated psychological and health consequences and decreased living standards.[35] In CMEA countries, the global economic crisis has been reflected in bottle-necks and problems of low productivity, rather than in capitalistic overproduction: with state-socialist employment legislation, the result has been underemployment rather than joblessness.

How should we regard work time? As an obstacle to human development, a sacrifice of free time that could be used for growth? Or as a possible terrain for directly extending individual well-being? This is not just a question for conceptual prescription, but also an empirical matter: whether work contributes to the well-being of individuals depends on the nature of the labour process in question, which in turn involves the social relations of production.

The great majority of workers in an industrial society are wage-labourers; they own no, or very few, means of production and have to sell their labour-power for subsistence and to be able to 'employ' raw materials and tools. The proportion of self-employed, largely in agriculture and retailing has been declining to low levels in most Western societies, and is typically low in the CMEA.[36]

What opportunities are there for human development during the time spent in wage-labour? Braverman (1974) forcefully argued that the twentieth century has seen a massive 'degradation of work', with skills being appropriated from the mass of labourers in the course of an increasing division of labour aimed at increasing control over the labour force for purposes of profitability.[37] While his account has a considerable impact, and certainly knocks the complacency out of many arguments of the 'post-industrial school' which sees work in general as becoming more rewarding with technological change, a general survey of these issues is required in addition to case studies, however telling the latter may be.

Recently some relevant analysis has become available that partly supports some of Braverman's theses while demonstrating that he tends to overemphasize the tendency towards deskilling in advanced capitalism.[38] Browning and Singelmann demonstrate that within the US there was a tendency for levels of more skilled jobs to decline within industries, but this was more than compensated in an economy-wide basis by the tertiarization of employment, the shift toward service, often white collar work.[39] Thus an overall increase in professional jobs has tended to reflect changes in the structure of the economy as a whole.

Such results pertain to other OECD countries, where the tertiarization of the economy is a fairly consistent phenomenon. (Whether processes are at all similar in the CMEA – despite the same general empirical tendency – awaits further study). Gershuny (1978) does report fairly similar results in a less detailed analysis for the UK, noting that it is important to consider the possible decreasing opportunities for the exercise of individual capacities within 'professional' occupations.[40] We find apparently similar trends in EEC countries too, but we should bear in mind the possibility that opportunities are decreasing for the exercise of individual capacities in professional or skilled occupations.[41] Spenner, however, on the basis of a study of job titles, concluded that over the 1970s upgrading of job content was the rule rather than the exception.[42]

Wright (1980) has conducted a study which estimates the proportion of the economically active population in the US that forms the working class under various definitions, using 1969 data.[43] While 88% of the economically active were wage earners, 52% were nonsupervisory wage earners (43% of men, 68% of women) and 42% were wage earners who were neither supervising nor semi-autonomous in their work (33% of men, 58% of women). According to Wright, large employers constitute less than 2% of the active population and small employers and petite bourgeoise some 11%; high management some 12%, low management and supervisors around 20%, and semi-autonomous employees (with some individual, though not collective, control over product design and work conditions) from 5% to 11%.

This data is relevant not only for considering opportunities for human development in work itself. Working hours, wages and fringe benefits are all related to class and occupational location, conditioning the free time that individuals have and the resources they may devote to using it. But our concern immediately is on the time spent within the labour process.

Insofar as the results of the studies reported above can be generalized to industrial societies as a whole – and there may well be significant differences between capitalist and state-socialist formations, as well as between 'central' and 'peripheral' industrial countries of both types – what can we say about the implications of this sixth to a quarter of adult time spent in paid work for human development?[44] First, that overall work appears to have been upgraded rather than deskilled over recent decades, although (i) deskilling is the norm for many jobs and the experience of many individuals, and (ii) there is reason to suspect that tertiary sector employment, the growth of which in the past offset sector deskilling, is currently increasingly subject to a process of rationalization itself. Many estimates of the introduction of new information technologies also forecast considerable deskilling and displacement of white-collar work.

The evidence on these matters is far from being conclusive and it is likely that substantial differences exist between industrial countries. For example, even within the OECD area, it has been argued that the peculiarly intense economic crisis facing Britain reflects an organization of capital and labour that has meant that workers have been unusually successful in preventing the reconstruction of the labour process along 'maximally efficient' lines.[45] The organization and experience of work in the CMEA countries is certainly different from that in the West, as evidenced in the lower productivity rates achieved even with similar technologies in the former.[46] We suspect that with certain exceptions these differences are more quantitative than qualitative, although the nature of state-socialism means that criticism and industrial action necessarily take on a form that is more political than is typical under capitalism.[47]

Second, we can venture some speculations concerning the determination of psychological development by these tendencies. Let us consider some recent pieces of American social research involving empirical, cross-sectional analysis. Kalleberg and Griffin (1980) report that Marxist class categories, occupational census titles, and job complexity/skill measures derived from the *Directory of Occupational Titles*, are each independently related to a measure of job 'fulfilment' based on six survey responses seeking information on the opportunities for growth, degree of challenge, and level of interest of one's work.[48] Employers and managers, for example, report greater fulfilment than do workers in addition to income differentiation, and within each class, occupations wielding more authority and demanding more skill also appear to provide more fulfilment. In terms of trends in human development at work, proletarianization, and the balance in favour of upgrading as opposed to deskilling, would appear to operate in different directions and their resolution is unclear.

Kohn (1976)[49] attempted to determine the relative correlation of two aspects of alienation – control over the product of labour (as yielded by ownership as against low hierarchical positions),[50] and control over the labour process (as indicated by the degree of supervision, and routinization and job complexity) – with subjective reports of powerlessness, normlessness and self-estrangement. These survey items relate to broad perceptions of one's life; a cavalier attitude to morality and the degree of influence one might have on social affairs or one's own life changes, little sense of worthwhile, goal-oriented living. The second aspect of alienation was found to be correlated with these factors among a sample of American men, and Kohn infers a causal process here (based on a job-history analysis) such that limiting occupational experiences are conducive to a perceived lack of self-direction in other areas of one's life. More recently, a longitudinal study by Kohn and Schooler, (1978) indicates that job and intellectual flexibility co-determine each other, so that psychological functioning can be restricted by, as well as itself limit, work experience.[51] Kohn's work provides perhaps the most convincing evidence to emerge from empirical sociology in support of the claim that psychological well-being is determined by what we do rather than by simply amassing more and more material possessions. Whether or not trends are observable here – and the data would seem to indicate a tendency for job complexity to increase through the postwar boom – two things stand out. First, that massive inequalities in the potential for psychological development are structured around the working conditions of industrial societies;[52] and, second, that this is condemning a vast proportion of our population to jobs that are unfulfilling both in terms of immediate experience and intellectual growth. If we add to these studies those of Levi which indicate that more than half of Swedish workers find their jobs either stressful or monotonous, and that stress is associated with psychological and health impairment, then the case for major transformation in the process of work becomes urgent.[53]

What sort of transformation here would further human development? Among the characteristics that must be cited we would include the opportunity for individuals to be very mobile between occupations, and to work for substantially shorter hours than is now the norm. Also there must be the opportunity for workers collectively and democratically to structure the conditions of work, and, where design skills are essential, to have access to such expertise both for evaluating the effects of working environments and for ergonomic design of whole production processes. These measures would constitute a radical transformation of the social relations of production: for them to form elements of a viable social order requires institutions which can replace the market – and the central planning of state socialism – as a means of co-ordinating the input to and outputs from different units of production and communities.

This is not to say that marginal improvements could not be achieved in all occupations in industrial societies. We would hope that such improvements are won not at a cost of defusing opposition to alienated labour, but rather that they will form part of the conditions within which awareness of the limitations of existing relations can be developed towards more effective pressure for change.

This applies equally to projects aimed at expanding the 'informal economy', and may be seen reflected in the different orientation of protagonists here. While a renaissance of simple commodity production, or a flowering of numerous small capitals is an unstable project (tending to evolve toward, or be subordinate to, monopoly capital or state industry),[54] there are doubtless opportunities for less alienated work here,[55] and it is possible for both prefigurative products and work forms to be developed by, for example, producer co-operatives.[56]

We are hesitant to pronounce on emergent trends in conventional formal work. It may be that an increasing quality of working life was more attainable during the postwar boom period, and that the current round of technological change under conditions of economic crisis will lead to an accelerated deskilling and even a polarization of the workforce, as the European Trade Union Institute, for example, argues.[57] Information-processing technologies could remove many boring and repetitive jobs. They can also be used to displace some skilled functions and render much white collar work subject to greater supervision and control. These technologies have considerable potential for improving the quality of working life, and certainly demand some new skills in production and operation: however, a major impact at present appears to be in the displacement of workers, especially female clerical staff. The evidence is far from adequate at present, but there are no grounds for assuming that a steady trend of job improvement for the majority of the workforce can be projected into the immediate future.[58]

We have yet to consider the unpaid employment largely carried out by women, housework. Housewives in the Szalai study reported more free time than did employed men, but the latter reported an even greater margin of free time when compared to employed women (who carry out housework on top of their waged work).[59] Both groups of women had considerably less free time than men on 'non working' days. Gershuny and Thomas reported that women in the UK and US alike were involved on the average in roughly six hours a day housework in the 1960s, but by the mid-1970s this had been reduced to something nearer four and a half hours (with a very slight increase in male housework).[60]

Domestic work subsumes numerous contradictory and conflictual tendencies. While on the one hand offering possibilities of self-determination ('Do It Yourself', etc.), it is on the other hand often a privatizing ritualized drudgery, receiving little social recognition or opportunity for human development. That the latter aspects seem to be dominant in industrial societies is suggested by Oakley's survey results revealing high levels of dissatisfaction with their tasks among housewives.[61] In American studies of the distribution of decision-making power within families in the 1950s, the wife's power was found to be well below the husband – except in respect of her own job and various household decisions.[62] So why is this not a realm of freedom, a household labour process performed for one's own perceived needs without unwanted supervision?

While housework does produce use-values, these tend to be aimed at the reproduction of living conditions rather than at new production; it is frequently monotonous 'maintainance' work like cleaning, for example (as is much women's paid employment). Unpaid and private (loneliness is common among working class housewives, especially those with young children), there is little opportunity for co-operative activity, and attempts to improve one's prospects or remuneration can imply conflict with the male 'breadwinner'. There is no obvious end to the work, and housewives' leisure time often contains, and is constricted by, a strong component of involuntary childcare. With the lack of obvious standards for the work,[63] recognition of one's labour becomes contigent upon interpersonal relations, and, with the sometime exception of cooking, is more likely to take the form of noticing shortcomings than of praise for a job well done.[64]

Finally we must consider unemployment as well as work. Unemployment levels increased drastically in OECD countries in the 1970s – from 2.6% to 5.4% of the labour force in the EEC from the early to the late 1970s, for example.[65] The severity of the problem varies, but growing concern is expressed throughout the OECD. (CMEA countries regulate their unemployment to a considerable extent, although some Eastern European planners now air the possibility of firms firing workers as a means of ensuring greater flexibility.)

A thorough review of studies of the effects of unemployment has recently become available.[66] Various macroquantitative studies have suggested links between stress-related pathologies (alcoholism, psychiatric illnesses, suicide) and unemployment rates in several English-speaking countries, but there is some controversy about the statistical bases of this work – and also about the role of intervening factors like family conflict. Unemployment has psychological and economic implications with different temporal structures, and some studies indicate long-term consequences for psychological development of unemployment.

Although unemployment may release the individual from an unrewarding job, and certainly allow free time from commitments, there are several reasons for expecting it generally to take on a regressive role with respect to human development in existing industrial societies. In a society where the wage nexus provides the main channel for a social valuation of the individual (as apart from that recognition that might be granted by friends and acquaintances), to find that the only commodity one can bring to the market, and one that is very close to the core of one's identity (capacity to labour), is not wanted by anyone can be a blow to self-esteem and sense of purpose. One is excluded from social production, and from the opportunities to interact co-operatively with other workers (and even to attempt to gain more control over the labour process through industrial action). Together with attacks in the mass media on the unemployed as 'scroungers', the inability to perform the role of wage-earner for one's family can also be a source of depression. Some unemployed workers may use their free time in a variety of growth-enhancing ways, but a considerable quantity of both anecdotal and systematic material indicates that human development is hindered by current forms of unemployment.

With this in mind, it is disturbing indeed to find that demographic changes, economic problems, and the possible applications of new technologies within industry are widely expected to result in continuing high levels of unemployment through the 1980s. Youth unemployment is particularly at issue here. While the political consequences of these developments ensure that attention will be paid increasingly to problems of the unemployed, we would contend that human development will not be served by treating them either as consumers of new leisure activities or as conscripts for new social or military services. These alternatives at best impose choice, at worse remove it, from the unemployed, whose psychological growth requires the exercise of agency individually and collectively.

Free time

Meeting the material needs of one's family and oneself through work and 'personal needs' activities leaves people with free time, which may be spent in leisure activities or otherwise. From Table 8.2, we see that the amount of free time that is available is typically related to sex and employment status. On working days, housewives would appear to have most free time; at weekends, men are advantaged over women. Large proportions of workers would prefer more free time to more money, according to surveys in several industrial countries, although this depends upon the trade-off between sufficient income and sufficient free time.[67] There has been some tendency for free time to increase, with shorter hours of paid labour and some labour-saving household technology.

Time budgets give us only a crude impression of free time activities. It appears from Table 8.2, for example, that an overwhelming proportion of free time is spent using the mass media: but we will be interested in the form of this activity, and whether it is performed as a solitary or communal activity, a shared or private experience. And what is the content of the media – television channels in different countries broadcast substantially different kinds of programmes, for example – and how is it attended to?[68]

Table 8.2 Free time

	Free time[1] as a proportion of:				Percentage of free time alloted to:										
	Twenty-four hour day	Waking hours			Education	Religion	Organization	Mass Media	Social	Conversation	Sport	Outdoors	Entertainment	Culture	Leisure/Travel	Resting
	Work days	Days off	Work days	Days												
Employed Men	14	35	21	56	6	1	2	43	14	6	1	7	2	0	7	8
Employed Women	10	25	15	40	4	2	2	40	15	6	1	5	2	1	7	9
(Housewives)	17	23[2]	25	30[2]	1	2	1	39	15	7	0	6	1	0	5	11

Source: Robinson, Converse and Szalai (1972) p. 131, Table 7, p. 132.
Notes: [1] Data refers to average of budgets, including those from Lima in Peru with industrial countries.
[2] Sundays.

The consumption of mass media in terms of the number of inputs available, appears to be typically much higher in industrial countries than elsewhere: more books, radios, televisions and newspapers circulate in these regions.[69] There are quite big national differences: a relatively large number of newspapers are consumed in the USSR per capita, the US broadcasts more television, and so on. Television has apparently become the dominant medium, the first source of news information for most people. The use of television has clearly been growing considerably, but patterns in different countries seem to vary. Thus Gershuny and Thomas found the growth in television use in Britain (averaging 20 minutes a day per person from 1961 to the mid-1970s) to be compensated for by a decline in radio use, so that with increasing leisure time, 'passive' leisure declined as a proportion of all leisure.[70] In the US, however, 'passive' leisure increased by some 10% as a proportion of leisure time, due to growing television use.

It hardly requires restating that the mass media are overwhelmingly owned by a few large concerns; that even when these are formally under public ownership, major policy decisions are made by senior state functionaries rather than by more democratic processes; and that the detailed organization of programming leads to the production of news and entertainment material that is strongly united toward the reproduction of hegemonic views.[71] But, in the West at least, some diversity of expressions is typical and people everywhere would seem to recognize that some critical stance is required in media consumption. It is, for example, widely reported that public opposition to the Vietnam war in the US was stoked by television coverage. In the face of government attempts to form opinion, television has surely widened people's awareness of what the world looks like, if not how it operates; and there are often educational broadcasts (usually for minority audiences) of a quality that few books and lectures can surpass.

It is easy to fall into what Enzenberger terms 'cultural archaism' in assessing electronic media: to fail to notice that they are accessible to considerably more people than traditional media, that media content can be influenced by counter-hegemonic forces, and so on.[72] But it is difficult to feel easy about the implications for psychological development of exposure to a regular diet of violence (such that American children typically have witnessed tens of thousands of fictional killings by the time they reach maturity) and social degradation (spiteful contests, etc.). There is an argument that television has helped to restore family life: husbands and wives spend more time together watching television instead of going their separate ways. But do they proceed to share their experiences, to use them to deepen their understanding of themselves and their relationship? Casual observation would suggest not, although the lack of communicative skills can hardly be blamed on the media.

Perhaps the term 'passive' leisure is not so inappropriate when applied to media consumption. Information is needed for action. Identification with other circumstances can be facilitated through electronic channels, but in what action does it result? Does increasing television viewing contribute to people's effectivity in their lives more generally?

It is easy to answer these questions negatively, on the grounds that private television viewing appears to be substituting for more active leisure or constructive agency, and that the amount of information transmitted means little if it cannot be put into practice. This may be because of its inappropriateness to most people's living conditions, its poor quality, or because there are features in the overall organization of the data that tend to lend to trivialization, instant forgetting, or to fatalistic attitudes. Impressionistic evidence here, however, needs to be amplified with more empirical research.

We would suggest that this research would tend to demonstrate that increasing exposure to mass media, beyond a certain point, itself related to the form and content of each medium, is counter-productive for human development. Two related characteristics may be cited here. First, that while each individual may process the material viewed in his or her own unique way, this does not make the programme any more of a dialogue: it remains a matter of the individual's choice between monologues. There is minimal scope for exercising one's capabilities other than those of private interpretation. Second, and related to the above, emotional experience is programmed by media controllers, and may even be of a diluted, desensitizing nature. Rather than using and developing one's capabilities for feeling in substantial human relationships, one is led into cycles of vicarious experience.

New information technologies could open up rather different possibilities. All that is required is the appropriate investment to create citizen access to data banks, to enable television viewers to contact both experts and other viewers electronically, and for community groups to prepare material cheaply for storage and transmission on cable television/computer networks. There are obvious advantages in this sort of scenario, although these may accrue more to those already rather more familiar with such technologies. Problems will be posed though: are controls to be individualistic; based on ownership or bureaucracy, or democratic? How are the costs to be set, and how are the advantages of wealth in producing persuasive professional material to be confronted? Studies of existing 'community access' television find rather similar ideological structures to be broadcast – a liberal consensus view of television and television staff as neutral leading to an exclusion of many diadvantaged groups, reliance on familiar official spokespeople, a surfeit of information and a lack of analysis, and adaptation to state controls.[73] Dialogue is still intrinsically limited and framed within dominant discourses, and ingenuity and pressure will both be needed to achieve more significant openness here.

We have so far ignored another major institution for transmitting information – or reproducing practices – which occupies a great deal of the time of most young people in industrial societies; that is, the educational system. As with the mass media, it is possible to criticize both the form of this institution (typically disciplinarian, hierarchical, based on a model of transmitting information rather than of developing critical learning skills) and the content of its teaching (which contains both material of great use and value and assemblages of ideology and ideologically structured facts).

We cannot possibly review the mass of research on education here: let us be content with a few general points. First, education is compulsory for younger students, although the degree to which alternatives to state and organized education are available, and the degree to which such alternatives constitute anything other than specialized training camps for the children of ruling classes and higher bureaucracies, is quite variable. Second, while education may be for some individuals a key to social mobility, for the population as a whole education is more of a response to the changing occupational structure (that itself determines the opportunities for mobility) than vice versa. Third, educational levels have increased rapidly in industrial societies since the war, along with the transformation of their economic structures, but massive structural inequalities by sex and class remain the norm, with schooling acting both to reproduce and legitimize the social division of labour.[74]

The same contradiction manifests itself. At the same time, schooling is a source of knowledge that can be used to increase individual and collective agency, and it imposes a form on this knowledge which fragments and mystifies and tends to prevent the exercise of this potential. The cultural systems of our social formations, like the political and economic ones, have acquired a fetishistic quality, and are seen as independent and even neutral structures rather than historical products of social relations.[75] But within all of them there is necessarily some potential for conflict and change, while in their functioning they do not succeed in rendering people into pre-programmed automata: the experiences they provide can be worked over, together with other experiences, to produce insights and forms of action that may transcend their bounds.

The state and the family

We have dealt with important social 'spaces' already in our discussion of human development in terms of social time. Under the headings above we have had to discuss the workplace (and thus the capitalistic or bureaucratic firm), and the media and educational systems. We have thus encountered some aspects of the economy and the political areas of industrial societies. Political life does not begin and end with the institutions outlined above, and although a small fraction of everyday life is normally consciously directed to this 'space' in the OECD and CMEA countries, it is important to consider the role of the state and the political area more generally in considering psychological potentials and growth. Another institution which forms a vital part of the medium within which we spend our lives, and whose presence is implicitly assumed in many of the categories of 'personal needs' and 'free time', is the family.

First, the realm of *politics*. Major decisions concerning the production and allocation of resources, the extent and nature of civil liberties and the relations that are to pertain between one group of people (a nation) and another are made, practically on a daily basis, in the boardrooms of business and the chambers of the state. The policies involved necessarily affect our daily life – the security of our work and the nature of its safety controls, the availability of health care and a welfare net, the freedom from surveillance and interference, the opportunity to meet with peoples of other nationalities and the very chances of war. It may seem strange that so little time is devoted to involvement in these decisions, but, making such decisions has become the work of select minorities; and their activity is the alienated will of the people in much the same way as the economy forms its alienated labour.

This alienation of agency is nowhere more clear than in the military threat that currently hangs over the head of all of us in industrial societies.[76] Martin Ryle has noted that 'as with the labour process, so in war too, the refinement of techniques reduces to almost nothing the role of human agency, which begins to reside solely in the will of those in charge', and asks whether the term 'democracy' is applicable to societies whose vast majority of citizens will have no say in decisions necessarily concerning their very life or death.[77]

The military branch of the state is typically organized in a far more secretive and hierarchically disciplinary structure than other branches. Even here there are significant differences among countries, with military conscription the norm in CMEA and some Western countries, with other Western countries relying on voluntary recruitment to the military (there has been a stepped-up recruitment drive in the current time of declining job opportunities). Of course, as with schooling (though perhaps less convincingly), it may be argued that the restriction of liberty involved in military service must be traded-off against the skills acquired; and again it must be noted that whatever skills of any worth are provided by such an experience are embedded in a framework of threat, national chauvinism and the like. Nevertheless, we might see rather more potential for self-activity in military arrangements stressing the role of civilian defence such as the Swiss and Yugoslav systems.

Turning to political life more generally, we can, with the advantages of hindsight, readily dismiss the idea that human development in industrial societies necessarily brings into being individuals who feel politically effective and satisfied with the operation of pluralist policies.[78] Both civil conflict in the 1970s and (in the West, at least) widely fluctuating levels of confidence in major political institutions and feelings of political powerlessness, demonstrate that these forecasts were ill-founded indeed.[79] But the very fact of fluctuating attitudes, seemingly concerned more with the current personalities' performance than with the structure of the institutions that select such leaders for such roles in the first place, suggests that this data has little direct bearing on the question of the contribution of the organized political system to human development.

A rather more revealing approach, which is also based on survey research, has unfortunately received rather less research effort: Mann has argued that, contrary to received sociological theory, the stability of liberal democracies does not rest upon a shared consensus of values about social goals, but rather that major ideological differences – which his data tends to show as structured along class lines – exist. But ideologies counter to dominant practices are typically poorly integrated and lack operational translations. Thus Mann cites studies indicating that among Americans, at least, a large proportion of the population (especially blacks and manual workers) tended to be in favour of collective action when specific questions were raised (e.g. about socialized health care), but to oppose it in favour of individual initiative when questions were posed in more general philosophical terms. It would appear that the existing political system confuses the attitudes of large proportions of the population.[80] To this clear constraint on their potential agency in securing desired goals must be added significant levels of misapprehension concerning the existing distribution of wealth and power, and the trends as per social equality, in Western societies. It may be that in state-socialism, with obvious exceptions, such 'false consciousness' is less prevalent: even if this is significantly the case, we doubt that it extends so far as to render human development and the trajectory of the political system consonant. A lack of perceived efficacy in political affairs goes hand in hand with a lack of development and integration of political affairs values.

But of course this state of affairs is not inevitably stable. If the state is fetished, then this fetishization of relations of power has to be continually recreated in the face of countertendencies of opposition and revolt. Our experience in the ballot box tells us that we exercise political power as isolated if 'equal' individuals.[81] The same fragmentation and subordination is imposed by the administrative and welfare apparatus. But nevertheless many forms of political opposition impel people towards collective action – as tenants, claimants, residents and (increasingly with state intervention), workers.[82] With the restructuring of the state in the current economic crisis, political conflict has developed around many areas of state activity, where the attempts to preserve welfare rights in the face of cutbacks have forced people to question the form in which these rights have been provided. The possibility of collective agency in aspects of social development which had previously been believed to be purely technical issues, is being debated in a significant minority of left organizations in, at least, some Western countries.

The *family* is an institution organized on quite different lines to the state, although, like it, a pre-modern institution has been taken up, transformed, and reproduced by capitalism and state-socialism alike. It is rather more obvious that large portions of time are spent within the family than it is that the same is effectively true of the state; for many daily activities are constituted as family activities. Furthermore, many people feel that they do not have enough time for family life – two out of three Western European parents would like to spend more time with their children (and practically the same number feel sorry that children now have less contact with their grandparents than previously),[83] although the fact that there are also many demands for more childcare facilities indicates that not all time spent in family contact is productive.[84]

The typical situation of the nuclear family in industrial societies appears to be an unusually contradictory one. On the one hand there are growing divorce and illegitimacy, and on the other hand there is a tendency for marriage ages to reduce and for divorcees to remarry (so that the situation has been described as being one indicating the rise of a new institution: sequential marriage).[85] What these tendencies signify is a matter of controversy, but they surely have implications for human development.

One way of interpreting the contradictory phenomena is to see the nuclear family as increasingly being the sole focus of meaningful relationships and emotional life in our societies.[86] True, the family has declined as a unit of production and become mainly a site of reproduction and consumption (although this is not to rule out the significance of existing family production, and the possibility of shifts towards greater domestic production of final goods).[87] Production relationships in industry have become less personal and more organized as commodity relationships (although, again, the growth of service work is to some extent a counter-tendency to this impersonalization); marriage offers apparently secure emotional and sexual relationships, economic partnerships and escape from the parental nest which, if scrutinized rather more closely, might reveal the security of marriage to be rather more limited). Small wonder that people rush into legal ties – and increasingly become disaffected, yet can typically find little alternative, and dismiss their problems as ones of the partner's inadequacy or the immaturity of both partners in selecting unsuitable companions.

Is all that is at stake too much haste in the search for emotional and economic security? This is undoubtedly a factor in explaining current developments, and has to be related to factors like the ideologies of leisure, mass consumption, and stereotypical sexuality, and the commercial exploitation and reification of generational differences, as well as to the structural changes in work (and other activities) mentioned above. But it might also be wondered whether there are not problems related to the composition of the nuclear family. Perhaps one intimate relationship is an insufficient source of feedback on one's actions and their deeper consequences for others – especially where this relationship will already be shaped along patriarchal lines. The development of the women's movement, of various forms of group encounter, of experimental family and household forms may reflect in part and help contribute to the overcoming of such problems.

Time and space constraints prevent us from going beyond this cursory examination of the family. Suffice to summarize our perspective here: existing family structures promote considerable pain and psychological rigidity in our societies – and so does their breakup, in the typical circumstances of insufficient individual resources and inadequate societal provisions for separation, let alone for the development of alternative family patterns.

Finally, it may be helpful to step back to consider the implications of structured social inequalities on interpersonal relationships. In an extensive review of empirical studies, Archibald marshals numerous studies from North America and Europe which relate alienation between people to class and status differences.[88] (His approach could readily be extended to gender disparities, although these have their own specificities). Such alienation here involves detachment and indifference in interpersonal relations, the pursuit of narrow and egoistic purposes, and the exercises of influence based on power differences, all of which are displayed across class and status boundaries – with corresponding emotions of dissatisfaction and hostility. Archibald views this in terms of the threats implicit in unequal social structures: on the one hand fears of resentment and rebellion, on the other, dangers of personal failure or the sanctions that the more privileged may exercise.

If we accept that a central component of human development is the ability to communicate and empathize with others, then it is clear that, in class and patriarchal societies, immense limits are placed upon such development on the basis of micro-stratification processes alone. Yet, as with so many of the phenomena touched on above, these are contradictory formations which carry the seeds of their own undermining. Divisions which inhibit trust and contact between social groups may provide opportunities for dominated, as well as dominant, groups to form their own distinctive analyses of their conditions. Class and sex solidarity among the exploited and oppressed are fragmented by economic, political and cultural institutions; nevertheless, it is here that the potential exists for the development of an understanding and exercise of collective human agency, for such development cannot take place out of the beneficiaries of social orders in which agency is alienated.

Likewise, the critical practice necessary to secure progressive change towards societies which can foster human development (as an explicit and central goal) actually requires the formation of organizations independent from dominant institutions. Whether these institutions obscure and fetishize real power relations, as in capitalist societies where free and fair exchange often seems to be the rule, or are more obviously instruments of domination, as in state-socialism where scientism forms a scanty cloak for bureaucratic privilege, they are primarily agents of control. In their technological scope, they distortedly mirror the growing capacities of human agency, behind which has lagged the development of the necessary democratic social forms. Pressure upon the institutions of rule must be complemented by the creation of alternative centres of power, if human agency is not once more to become the power exercised by humans over and against humanity.

Taking stock

We have only scratched the surface of the problem of human development in industrial societies. What is outstandingly clear, even in the crude and limited set of indicators employed here, is that human development processes are exceedingly complex and contradictory. If human development is to be the goal of development, then in no way can industrial societies be said to be 'developed'. In an effort to summarize some of the tendencies and trends of most relevance to the future course of human development, Table 8.3 gives an impressionistic picture of trajectories that appear to be dominant: in most cases, in both capitalist and state-socialist societies.[89]

The Table perhaps provides too gloomy a view. Resistance to oppressive directions of development repeatedly springs up, in diverse forms. Frequently this is temporarily contained by superficial modulations of the main trend. But new poles of resistance and new strategies for change are arising, and there is reason to hope that diverse struggles can be related together in new and effective ways. The peace movements of the early 1980s, for example, raise questions of the democratic control of the state (foreign policy, alternative defence strategies, control over party bureaucracies), of the economy (the arms trade, socially useful conversion of military industries), and of internationalism (links between popular movements in the international state system). With these, solidarity of peoples across the world to preserve a natural environment and create social systems within which development may be human-centred becomes a key goal. If the world is inching closer to ultimate disaster, if many certitudes of the 'good life' are crumbling, then it is nevertheless true that the possibility of human development is shining more brightly. It is vital that this possibility should be glimpsed by many more people. This is not to start resistance to maldevelopment, for resistance is already continuous, and usually takes maldeveloped forms itself. If we are to act to exercise our agency, we need to grasp the contradictions of everyday life to both understand the world and to change it. Understanding what human development could be, can help us recognize what it is.

Table 8.3

Political Sphere   Long-run Tendencies
  (1) In response to popular-democratic movements, greater enfranchisement, devolution of some powers; but, associated with this, removal of critical issues from parliamentary arena to more executive and corporatist policy making. Waning confidence in institutions.
  (2) Development from securing conditions of production, to support for declining sectors and provision of infrastructure, to active intervention in restructuring productive capital. Waning of legitimization based on free market, politicization of economy. Danger of fiscal crisis.
  (3) Growing capacities for surveillance and control of populace, growing capacities for military destruction of ecosphere. Opposition to these tendencies through civil liberties and disarmament movements.

Conjunctural Tendencies
  (1) Crystallization together of anti-bureaucratic and anti-inflation attitudes to provide populist support for restructuring of welfare state. Provokes interpenetration of political and economic defensive movements, with some development of alternatives to technocracy, and affirmative conceptions of action.
  (2) Growing threat of military confrontation: provokes reassessment of democratic and internationalist practices.
  (3) Breakdown of social democratic consensus in face of own contradictions: political polarization, fragmented by failure to cope with new interest groups and co-optational practices.

Economic Sphere   Long-run Tendencies
  (1) Growth of large corporations as dynamic economic actors, with transnationalization of investment and control. Perception of increased remoteness of policy makers from affected workforce, environment and communities.
  (2) Gradual increase in scope and scale of working-class organization, with capacity to take major industrial action somewhat offset and channelled by non-democratic, bureaucraticized modes of organization.
  (3) Technological development increasing productivity and levels of capital investment required; major shifts in occupational structure, in need for planning and infrastructural development. Changed experience of labour process, skills and autonomy required, more awareness of ecological and technological issues.

Conjunctural Tendencies
  (1) Contraction of employment, stagflation and increased competition for resources and markets in wake of postwar boom. Major sources of psychological damage, and of increased awareness of conflicting interests.

Table 8.3 *Continued*

Economic Sphere   Long-run Tendencies   continued
(2) Attempted state regulation of industrial conflict, and associated politicization of economic demands.
(3) Deskilling and displacement of traditionally technical and white-collar occupations.

Cultural Sphere   Long-run Tendencies
(1) Development of means of communication capable of cheaper transmission of more information, and of more flexible use. Application in ideological domination, but significant growth of alternative facilities and conflict within media, and of cynical attitudes in general public.
(2) Mass education and exposure of population to wider range of modern science and culture, though these often torn from social roots and decadent. Increased understanding of technical organization of production.
(3) Mass consumption society developing, with attempted commoditization as cultural goods of wide range of human activities, or introduction of consumer goods into existing practices. Reactions against depersonalization, mechanization. Some growth of counter cultures.

Conjunctural Tendencies
(1) Tightening of control over media, retreat of entertainment into spectacle and horror.
(2) Retrenchment of alternative ways of life groups as retreatists, marginals (sometimes making a good living in view of fear concerning health, etc.) or more conventional activists.
(3) Use of new technology to mobilize political support by neoconservative groups in US . . . and elsewhere?

Family Sphere   Long-run Tendencies
(1) Displacement of major productive activities of family of commodity production. Family as unit of consumption wooed and glorified by market, yielding high aspirations for privatized leisure time. Involvement of women in employment strengthens economic position, enables extended fight for equality.
(2) Displacement of major reproductive activities of family by state (education, welfare), and growing 'policing of families' by professionals and para-professionals. Decline of extended family ties and isolation of generations; pressure for improved childcare facilities.
(3) Increasing stress on family as haven of heartless world: growth in marriage break-up, in tranquillization of 'unsuccessful' housewives, in counselling services, in extra-marital sexual activity. Search for alternative family structures.

Conjunctural Tendencies
(1) Resurgence of familialist movements in synchrony with pressure to expel women from labour force and strengthen discipline of youth. Attempts to reverse gains of women's movement.
(2) Conflict around 'cuts' in services for women, children, parents and emergency of action to democratize and make such services non-sexist.
(3) Upsurge of religious cults, and their transformation from eccentric fringe activities to well-organized business operations yielding communal experiences.

A provisional assessment

We are not quite in the situation of Jeddah's radio service in January 1979, which was unable to provide a weather report or forecast because the airport which supplied these reports was closed due to the weather.[90] Our reports and forecasts must remain provisional, however, for the uncertainty and turbulence of social affairs is so great as to defy ready conclusions. We may note alarming trends and a deepening of exploitation in productivity, psychologically and in other ways. But we may note new areas of resistance, some oriented to constructing new forms of action as well as defending traditional practices.

In Table 8.3 we attempt to outline some major tendencies which frame the circumstances of human development in Western societies. (Our experience of CMEA countries is limited, but suggests that a fair proportion of these notes would need to be changed to cover the situation there.) We have distinguished between long-term tendencies and conjunctural ones reflecting the immediate period of restructuring, but do not wish to obscure the essential unity of the two, or the ground of both in 'unmastered human practice' rather than in some super-historical laws.[91]

In Table 8.4 we go on to consider some of the points of criticism, action and conflict that emerge out of this apparent chaos. Here we focus especially upon different possible strategies, and it may be noted that Anderson's three types of agency can be 'mapped' onto the four types of strategy, with the first two strategies involving attempts to exercise agency over the conditions and consequences of one's immediate way of life, the third requiring collective action to acquire more agency within the social order, and the fourth seeking to apply to the construction of a radically different social formation.

Again we should note that these are not exclusive strategies: that often an individual must make use of all them, and that on a societal level it is essential that aspects of all four be developed if a world in which human development can be the cornerstone of all development processes can be secured.

Table 8.4 *Strategies for social change*

		Social responsibility	Counter culture	Reformist	Contentary
Social Sector	Dominant aspects criteriarized as exploitative, oppressive, etc.	Involves the exercise of individual discrimination, consumer choice, whistle-blowing. Can be expensive to adopt different WOL here, so most possible for those with most resources. Exposé work involves dangers to career, legal position etc. Can have demonstration effect if publicized; those in influential positions may exercise more influence. Demands individual commitment and unless reformist-style organization established, liable to wane with individual's fortunes.	Involves living AWL through new institutions, through becoming part of a (hopefully vanguard) minority. Demands resources to the extent that continued viability does not rest on manifold dependency on DWL. By virtue of minority, and affluent constitution, effects likely to be limited: demonstration may be effective, but innovation may be siphoned off for commercial exploitation. In more expensive climates, may be victimized. Dangers of ghettoisation, or of internal pressures due to felt need to present impossibly good public image.	Involves using organized forms to seek to influence dominant institutions in progressive ways, either through pressure-groups or by mobilizing public opinion. Requires channels of access to wide public (media) *or* to technical experts and bureaucrats. Liable to be confined within limits of reform, to be bureaucratized, incorporated, to foster illusions in evolutionary progress of DWL. But may also gain real improvements (if always open to erosion), tactical advantages, and increased sense of efficiency and commitment of members.	Involves attempt to build oppositional institutions based on shift in social power. This requires mass support if not to become hopelessly purist and fanatical. Runs risk of losing internal democracy in face of isolation and repression, of physical danger through conflict with authority. Also liable to opportunist seizing of issues and fragmentation of struggles. But a vital component of moments of upsurge, offering possibilities for establishing new bases of power, for exposing structural roots of problems, etc.
Private capital	Private ownership of means of production. Profit-guided via market allocation. Workers just one 'factor' of production. Large-scale monopoly organizations.	Support small firms and shops, products designed as ecologically sound and healthy, socially responsible organizations. Join consumer associations and pressure groups. 'Do-it-yourself'.	Seek work in co-operatives, communal production, etc. Establish 'alternative' shops, consume 'alternative' products. Strive for self-sufficiency.	Seek legal controls over work conditions, and product safety and environmental impact. Nationalization of vital industries, or those failing to serve social needs. Education of consumers.	Organize for alternative planning of major industries based on workers' plans, seizure of control away from monopolists.

2 State sector	Large bureaucracies, restricted access to decisions, hierarchical patterns of control, operate with notions of welfare geared to requirements of Sector 1 for trained, healthy but docile labour force and consumers.	Exercise choice to favour more responsive smaller organization, to preserve services that are in decline.	Seek alternative services; set up 'free' schools, 'people's clinics', 'crash pads', new information services, etc.	Seek to integrate counter culture alternatives into mainstream. Press for more participation of community groups in governing bodies. Seek democratic reform of state agencies.	Mobilize groups at the 'receiving end' (school children, claimants, invalids, etc.) together with more proletarianized state employees to demand new priorities, new controls.
3 Family sector	Privatized production, often tied to petit bourgeois 'community' informal sector or patriarchal family production. Restricted to 'marginal' needs, and slow to respond to changes in family structures and social relations.	Attempt to make own family relations more open, more equitable.	Live in new family forms; communes, etc. Encourage new forms of social encounter – personal growth movement, etc.	Press for socialized childcare, recognition of value of domestic labour. Support campaigns against parental assault on children, for shelters for runaway children and battered wives.	Support militant groups seeking to erode patriarchal system (women's movements, men against sexism).

Notes

1. Data cited from P. Barker, 'Whistling in the dark: social attitudes as we enter the 80s', *New Society*, 29 November 1979, pp. 480-6.
2. Source: Smith G. and Kellner P., 'The Good, The Bad and The British', *The Sunday Times*, 1980, reporting on Mori poll.
3. Gallup's study is available in 18 volumes, but my source is G. Gallup's report in *Readers Digest*, October 1976. Earlier work tending to show correlations between level of economic development and self-reported satisfaction include H. Cantril, *The Pattern of Human Concerns* (Rutgers University Press, New Brunswick, 1965); a discussion of such work, together with the 'structural regularity' that within countries not undergoing revolutionary change, the poor are less satisfied than the rich, on the whole, see my discussion in Encel, S., Marstrand, P., Page W. (eds.) *The Art of Anticipation*, London; Martin Robertson, 1975.
4. Cf. Glyn A. and Harrison J., *The British Economic Disaster*, London: Pluto, 1980.
5. Ornaur H., *et al.*, *Images of the World in the Year 2000*, The Hague: Mouton, 1976.
6. *The Europeans and their Children*, Brussels: EEC, 1980.
7. Sources: Gallup and Harris surveys reported in T. Wicher, 'The US Satisfaction Boom', *International Herald Tribune*, 20 February 1978; J. Rogaly, 'The American's Growing Pessimism', *Financial Times*, 8 August 1978.
8. Research by John Pollack reported in Raab, S., 'Pervasive Face of Crime seen changing US life', *International Herald Tribune*, 19 September 1980.
9. E.g. Paris-Match poll reported in M. Blune, 'Uncorking a Nuclear Bomb Shelter', *International Herald Tribune*, 27 February 1980.
10. But this fails to cope with a trend reported by Barker op. cit.: between 1959 and 1980 there were large decreases in the proportion agreeing that most people are inclined to help others, and similarly increases between 1967 and 1980 in the proportion agreeing that the future is so uncertain that the best thing to do is to take each day as it comes.
11. E.g. Hall P. *et al.*, *Change, Choice and Conflict in Social Policy*, London: Heinemann, 1975; Miliband, R., *The State in Capitalist Society* London; Weidenfeld and Nicholson, 1969; Cockburn, C., *The Local State* London; Pluto, for British case studies, 1978. For an interesting American analysis of local authorities, see Whitt, J. A., 'Toward a Class-Dialectical Model of Power', *American Sociological Review*, 44, 1979, pp. 81-100.
12. The Alternative Ways of Life Project, co-ordinated by M. Wemegah.
13. These issues are detailed in Miles I. and Irvine J. (eds.) *The Poverty of Progress*, Oxford: Pergamon Press, 1982.
14. A study not covered in the book in question suggests that some 20% of North Americans have diagnosable mental disturbances: another 25% suffer severe emotional stress; some 15% receive mental health treatment per year; in industrial countries of North America and Europe over 16% of the population have psychiatric problems observably impairing their lives; 15% of US children 'need help for psychological disorders'; and so on. Source: R. D. Lyons, 'Mental Health in the US', *International Herald Tribune*, 17 September 1971, reporting on President's commission in

Mental Health. Of course, professionals in such fields have everything to gain by advocating broad definitions of the problem under their ambit.
15 It has to be stressed that this is a preliminary study: the review of literature has been far from thorough and largely restricted to English-language sources. We have sought to avoid duplicating work published by the Alternative Ways of Life Project, although it would be possible to take quite a large proportion of that work and reinterpret it for present purposes.
16 See I. Miles, 'A Chain Reaction?' (1981) presented at Vision of Desirable Societies – III conference at CEESTM, Mexico City, to be published in a volume edited by E. Masini; and on related themes Brown, B., *Marx, Freud, and the Critique of Everyday Life*, New York: Monthly Review, 1973; Frankl, G., *The Failure of the Sexual Revolution*, London: New English Library, 1975; Reiche, T., *Sexuality and Class Struggle*, London: New Left Books, 1970.
17 Nudler, O. and Mallman, C., 'A Preliminary Integration Proposal' to the GPID project (Bariloche Foundation, 1980).
18 See, in particular, de Charms, R., *Personal Causation*, New York: Academic Press, 1968.
19 Perry Anderson, *Arguments within English Marxism*, London: New Left Books, 1980.
20 Anderson has earlier cited E. P. Thompson's definition of history as 'Unmastered human practice' from *The Poverty of Theory*, London: Merlin, 1978.
21 See other contributions to this volume, and I. Miles, *Social Indicators for Human Development*, London: Frances Pinter, 1985.
22 Szalai A. (ed.), *The Use of Time*, The Hague: Mouton, 1972.
23 E.g. Witkin, H. and Lewis, H. (eds.), *Experimental Studies of Dreaming*, New York: Random House, 1967. See also Fromm, E., *The Forgotten Language*, New York: Grove Press, 1957.
24 Timpanaro, S., *The Freudian Slip*, London: New Left Books, 1976, original publication *Il Labsus Freudiano*, La Nuova Italian, 1974.
25 Jacoby, R., *Social Amnesia: a Critique of Conformist Psychology from Adler to Laing*, Boston: Beacon Press, 1975. See also Fromm, E. *The Crisis of Psychoanalysis*, Harmondsworth: Penguin, 1978; the Radical Therapist Collective, *The Radical Therapist*, Harmondsworth: Penguin, 1974.
26 Jay Gershuny has uncovered a large collection of diaries kept in Britain earlier this century in which dreams were recalled by participants in the study. He suggests an analysis aimed at relating dream content to major social and political events of each day, as well as to individual characteristics. But what might be more interesting would be an attempt to assess whether the mere act of dream recall and description regularly carried out, had effects on the psychology of the individuals involved.
27 E.g. Stewart, K., 'Dream Theory in Malaya' in Tart C. (ed.) *Altered States of Consciousness*, New York: Anchor Books, 1972; Lee S. G., 'Social Influences in Zulu Dreaming', *Journal of Social Psychology*, 47, 1958, pp. 265-283.
28 Marcuse, H. *Eros and Civilization*, London: Abacus, 1972.
29 Gershuny, J. I. and Thomas, G. S. *Changing Patterns of Time Use*, SPRU Occasional Paper, No. 13, Science Policy Research Unit, Falmer, 1980.

30 E.g. Orbach, S. *Fat is a Feminist Issue*, London: Wildwood House, 1978; and on cigarette use see Jacobson, B., *The Ladykillers: why smoking is a feminist issue*, London: Pluto, 1981; for AWL perspectives see contribution to Miles, I. Irvine J. (eds.) *The Poverty of Progress*, Oxford: Pergamon, forthcoming, and BSSRS *Our Daily Bread*, London, 1978. See also the journal *Politics and Food*.
31 Fast-food and convenience foods are growing in popularity at the same time as whole food and health food, for example in Western countries.
32 Calculations are based on ILO data, and it is noteworthy that smaller proportions of this age-range are in paid labour (including family labour) in Third World countries.
33 Cacace, R., *Employment and Occupations in Europe in the 1980's*, Strasbourg: Council for Cultural Cooperation, 1980.
34 E.g. Gershuny and Thomas op. cit., for the UK and US, where declines of some 5-6% in hours of paid work over a recent decade are noted.
35 See Jahoda, M. and Rush, H., *Work, Employment and Unemployment*, SPRU Occasional Paper No. 12, Science Policy Research Unit, Falmer, Sussex, 1980.
36 *The Role of the Tertiary Sector in Regional Policy: Comparative Analysis* (EEC, Brussels); Downing H.C. and Singlemann, J. 'The Transformation of the US Labour Force', *Politics and Society*, **8** (3-4), 1978, pp. 481-509; Bechhofer F. and Elliot, D. 'Persistence and Change: the Petit Bourgeoisie in Industrial Society', *Archives Europeen de Sociologie*, **16**, 1976, pp. 74-99.
37 H. Braverman, *Labour and Monopoly Capital*, New York: Monthly Review Press, 1974.
38 For arguments with Braverman that are more theoretical or case-study oriented, see Burawoy, M. 'Towards a Marxist Theory of the Labour Process', *Politics and Society*, 8, (3-4), 1978; studies in the *Cambridge Journal of Economics*, Vol. 4, September 1980, and the introduction to them by Elburn *et al.*; T. Elger, 'Valorization and Deskilling', *Capital and Class*, No. 7, 1979, pp. 58-99.
39 Browning and Singlemann, 1978, op. cit. See also Wright, E. O. *Class Structure and Income Determination*, New York: Academic Press, 1979 for an account of similar results from a follow-up study focusing on the degree of worker's autonomy in the labour process.
40 Gershuny, J. I., *After Industrial Society?*, London: Macmillan, 1978.
41 Gershuny, J. I. and Miles, I., Service Employment: Trends and Prospects (EEC: Fast Project Working Paper No. 4, Brussels, 1982); Gershuny J. I. and Miles, I., *The New Service Economy*, London: Frances Pinter, 1983.
42 Spenner, R. L., 'Temporal Changes in Work Content', *American Sociological Review*, **44**, 1979, pp. 968-975.
43 Wright, E. O., 'Varieties of Marxist Conception of Class Structure', *Politics and Society*, **9** (3), 1980, pp. 323-370.
44 Thus Paci, M., 'Class Structure in Italian Society', *Archives Europeens de Sociologie*, **20**, 1979, pp. 40-55 argues that the social structure associated with Italian industry reflects, not economic backwardness, but a specialization of Italy within the EEC in production for richer European regions on commodities for which small firm organization is functional.
45 Kilpatrick, A. and Lawson, T. 'On the Nature of Industrial Decline in the

U.K.', *Cambridge Journal of Economics*, 4, 1980, pp. 85-102.
46 See Haraszti, M. *A Worker in a Worker's State*, Harmondsworth: Penguin, 1977, for an account which stresses the alienated quality of labour in Hungary.
47 See Hulubenko, M. 'The Soviet Working Class: Dissent and Opposition' in *Critique*, No. 4, 1975, pp. 5-26.
48 Kalleberg, A. C. and Griffin, C. J. 'Class, Occupation and Inequality in Job Rewards', *American Journal of Sociology*, 85 (4), 1980, pp. 731-768.
49 Kohn, M. C. 'Occupational Structure and Alienation', *American Journal of Sociology* 82 (1), 1976, pp. 111-130.
50 Although the regulation of production by market signals may itself be seen as a source of structural alienation.
51 Kohn M. C. and Schooler, C. 'The Reciprocal Effects of the Substantive Complexity of Work and Intellectual Flexibility', *American Journal of Sociology*, 84 (1), 1978, pp. 24-52.
52 In addition to those derived from the economic rewards for different forms of work and property ownership. As national studies in Szalai (1972) op. cit. show, too, the less pleasant and growthful jobs are those with the longest work hours.
53 E.g. Levi, L. *Quality of the Working Environment*, reports from the Laboratory for Clinical Stress Research, Stockholm: Karolinska Institute 1978.
54 Cf. Miles, I., 'New Technology, Old Orders' in Masini, E. (ed.), *Visions of Desirable Societies*, Oxford: Pergamon, 1983.
55 Often preferable to the psychological costs of unemployment (Jahoda and Rush, op. cit.), for while work may not fulfil, devaluation of one's capacity to contribute to social production is liable to directly undermine self-esteem.
56 For the co-optation of co-operatives by existing institutions, and strategies against such tendencies, see *Undercurrents* No. 41 (August-September 1980), where a number of case studies and general issues are discussed. On political problems of the informal economy, see Henry, S., 'The Working Unemployed', *The Sociological Review*, 1982, Vol. 30, No. 4, pp. 460-77.
57 European Trade Union Institute, *The Impact of Microelectronics on Employment*, Brussels: ETVI, 1979. See also Miles (1983) op. cit.
58 It should also be noted that the ability of workers to find preferable employment is severely circumscribed: viz. R. Blackburn and M. Mann, *The Working Class in the Labour Market*, London: Macmillan, 1978 (who also point out that the average worker finds driving the car to work rather more skill-demanding than their daily job).
59 Szalai, op. cit. See Table 9.2.
60 Gershuny and Thomas, op. cit.
61 Oakley, A., *The Sociology of Housework*, London: Martin Robertson, 1974.
62 See Gillespie, D. L. in Dreitzel, H. P. (ed.) *Family, Marriage and the Struggle of the Sexes*, New York: Macmillan, 1972.
63 Although various commercial and institutional (e.g. social services) pressures do set standards: imagine the consequences of a conservative-backed 'wages for housework' in this respect!
64 Thanks to Chris Zmroscek for this point. For discussion of housework, see

Eva Kaluskaya, 'Wiping the Floor with Theory, *Feminist Review*, No 12, 1980, pp. 27–54; M. Molyneux, 'Beyond the Domestic Labour Debate', *New Left Review*, No 116, 1979; A. Foreman, *Femininity as Alienation*, London: Pluto, 1977. For empirical analysis see Oakley, op. cit.

65 *European Economy* special issue 1979: 'Changes in Industrial Structure in the European Economies Since the Oil Crisis'.

66 Jahoda and Rush (1980), op. cit. Miles, I., *Adaptation to Unemployment?*, SPRU Occasional Paper No. 20, Science Policy Research Unit, Falmer Sussex, 1983.

67 E.g. Katona, G. Strumpel, B. and Zahn, E., *Aspirations and Affluence*, New York: McGraw-Hill, 1971; MORI opinion poll reported in *New Statesman*, 6 November 1980. In Third World countries, and CMEA ones, longer hours are worked than in the OECD area, on average (Cole and Miles, op. cit.).

68 See Williams, R. *Television: Technology and Cultural Form*, London: Fontana, 1974, for a comparison of UK and USA stations in terms of their programming, situated within an incisive study of the shaping and consequences of this particular medium.

69 Cf. Cherry, C. *World Communication*, New York: Wiley, 1971, Harms, L. S. *Human Communication*, New York: Harper and Row, 1974.

70 Op. cit.

71 For British data on media ownership and control see McShane, D., *Using the Media*, London: Pluto, 1979, Tunstall, J., *The Media are American*, London, 1978. For quantitation analysis of news content, see the Glasgow University Media Group, *Bad News*, London: Routledge & Kegan Paul, 1976 and *More Bad News*, 1980; *Really Bad News*, London: Writers and Readers, 1982. For an approach to the 'hidden curriculum' of television programmes see Gerbner's work: Gerbner, G. *et al.*, *Communications Technology and Social Policy*, New York: Wiley, 1977; Gerbner, 1970, 'Cultural Indicators', *Annals of the American Academy of Political and Social Science*, 388, 1970, pp. 69–81.

72 Enzenberger, M. M., *Raids and Reconstructions*, London: Pluto, 1976.

73 Bibby, A. *et al.*, *Local Television: Piped Dreams?*, Milton Keynes: Redwing Press, 1979.

74 For a vast literature, a varied selection: Halsey, A. H. *et al.*, *Origins and Destinations*, Oxford: Clarendon, 1980; Dale, R. *et al.* (eds.) *Schooling and Capitalism*, London: Routledge & Kegan Paul, 1979; Bowles, S. and Gintis, H., *Schooling in Capitalist America*, New York: Basic Books, 1976; Castles, S. and Wustenberg, W. *The Education of the Future*, London: Pluto, 1976 (includes accounts of state-socialist educational practice).

75 Cf. Johnson, C., 'The Problem of Reformism and Marx's Theory of Fetishism', *New Left Review*, 119, 1980, pp. 70–96; D. Wells, *Marxism and the Modern State*, Hassocks, Sussex: Harvester Press, 1981; Jessop, B., *The Capitalist State*, London: Martin Robertson, 1982; Cleaver, H., *Reading Capital Politically*, Hassocks: Harvester Press, 1982; Ollman, B., *Alienation*, Cambridge University Press, 1976.

76 This threatens any development, human or otherwise – at a massive antimissile march in London recently, one strangely garbed and made-up individual carried a banner with a slogan implicitly testifying to this:

'Mutants Against the Missiles'. Then there were the group demanding 'Historians for the Right to Work: ensure a continuing supply of history'. As noted above, significant minorities – in some cases in 1982, majorities – in industrial countries believe that nuclear war is more likely than not.

77 From 'Nuclear Disarmament: Democracy and Internationalism', mimeo: Brighton, 1980; submitted to *Radical Philosophy*. See his *The Politics of Nuclear Disarmament*, London: Pluto, 1981.
78 Lane, R. E., *Political Man*, Glencoe, Illinois: Free Press, 1972.
79 For example, confidence in major institutions like the legislature, presidency, judiciary and so on reached new lows in the US during Watergate, but has increased since: in 1978 confidence in political institutions were held by less than 70% of the American population – data from Harris Poll cited by A. Mitchell, *Social Change: Implications of Trends in Values and Life-styles*, Values–Lifestyles programme 1979. In the UK in late 1977 less than 50% of people surveyed felt any institution was doing a good job (Kellner, P. 'Who Runs Britain?', *Sunday Times*, 18 September 1977).
80 Mann, M., 'The Social Cohesion of Liberal Democracy', *American Sociological Review*, 35, 1970, pp. 423–439. Mann also cites some British data of more limited scope in support of his thesis.
81 An experience which seems to be on the decline where voting is not compulsory. In the 1980 US election only 52% of those eligible voted; the two main candidates thus between them could only muster the support of a minority of enfranchised Americans!
82 See London-Edinburgh Weekend Return Group, *In and Against the State*, London: Pluto, 1980; CSE State Group, *Struggle over the State*, London: CSE Books, 1979; Holloway, J. and Picciotto, J. (eds.) *State and Capital*, London: Arnold, 1978.
83 EEC, *The Europeans and their Children*, Brussels: EEC, 1980.
84 It should also be noted that the absence of adequate day care facilities means that many children of working parents are left 'unsupervised' for long periods: perhaps 20% of 5–10 year-olds in Britain according to Simpson, R., *Day Care for School Age Children*, Manchester: Equal Opportunities Commission, 1978.
85 There appears to be few exceptions to the tendency for divorce rates to rise, although those rates vary tremendously; cf. Gibson, C., 'Social Trends in Divorce', *New Society*, 5–7, 1973 pp. 6–8. The only large drop in divorce rates cited by Gibson is Rumania, following changes in family law. Some 40% of US marriages now seem to end in divorce, with 80% of divorcees remarrying: data cited in Ehrenreich, B. and English, D., *For Her Own Good*, London: Pluto, 1979.
86 Viz. Dreitsel, H. P., op. cit.; Poster, M., *Critical Theory of The Family*, London: Pluto, 1978: Gordon, M., *The Nuclear Family in Crisis*, New York: Harper and Row, 1972; Barret M. and McIntosh, M., *The Anti-Social Family*, London: New Left Books, 1982.
87 For example Gershuny, J. I. *After Industrial Society?* London: Macmillan, 1978; Kumar, K., *Prophecy and Progress*, London: Penguin, 1978.
88 See, in addition, Miles and Irvine, 1982, op. cit., and Fischer, D., *Major*

Global Trends and Causal Interaction Among Them, Tokyo: United Nations University (HDRGPID-76/UNUP-341), 1981, for a different, but not contradictory, assessment.
89 Archibald, W. P., 'Face-to-Face. The Alienating Effects of Class, Status and Power Divisions', *American Sociological Review*, 41(5), 1976, pp. 819–837. For an interesting depth study, see Sennett R., and Cobb, J., *The Hidden Injuries of Class*, Cambridge University Press, 1972.
90 Pile, S. *The Books of Heroic Failures*, London: Futura, 1980.
91 For a lengthier account of much the same set of issues, see I. Miles, 'Social Choice and Life Styles in Industrial Countries', *International Development Review*, 22 (2/3), 1980, pp. 67–71.

9
Human Development and Childhood

Eleonora Barbieri Masini

1 Human development emerges from needs-social character dynamics: a theoretical tentative basis

E. Fromm says that only through a 'picture of the social character, tentative or incomplete as it may be, do we have a basis on which to judge the mental health and sanity of modern man'.[1] He goes on to say that the 'social character is the nucleus of the character structure which is shared by most men of the same culture in contrast to the individual character in which people belonging to the same culture differ from all others'.[2]

What is important is that the social character is functional to a specific social system. The values of a specific society internalized by the members of the same society build up the social character which, as such, moulds the people in a given social system to make them function well in that society, and the society in its turn will go on functioning and existing.

As a consequence, the content of the social character is determined by the society, by its values and its norms and the role of the person in that specific society emerges.

Here I would like to go beyond E. Fromm's use of the social character concept and relate it to every social system. He talks in fact of culture and of society in general terms. Each social system has its values and norms which are internalized by its members, but as each person belongs at the same time to different social systems, he or she will internalize the different (or sometimes the same) norms and values. A man belonging to Europe, considered as a social system, will internalize the basic norms and values related to industrialization, to economic priorities, to freedom of choice, to participation in political decisions (these, of course, are generalized values, for argument's sake; differentiations are not included).

The same person could also belong to a specific country (social system) such as France, where the meaning of belonging to it has a very strong value which, internalized, differentiates the social character of the Frenchman from that of other Europeans. The same person could also internalize the values of the South of France, such as the religious ones related to Catholicism, if he lived in that region.

I shall go on a little further and consider, though differently from E. Fromm, the family as a social system which has a social character of its own related also to its surviving and functioning. It may be the case of the authoritarian value as central to the role of the father (family) and, as a consequence, central to the social character of that specific society. Such a central value will influence the roles in the entire family. The family in this case is the internal circle of all the other concentric social systems with specific and sometimes different social characters of the various members.

At the same time the family transmits to the child the social character which the parent lives in the different social systems to which that parent belongs. The methods of transmitting (educating) are tied to the same social character lived by the parents and transmitted to the children. It is hence a matter of content as well as of methods of transfer of social character, values, norms.

Along with different social characters, each person has individual needs which are biophysical, psychosocial and spiritual. Such needs change their priority according to their satisfaction which is offered by the society.[3] The society in turn offers satisfaction to needs related to its social character based on the norms and values of that specific society.

This point is important because in a society where the social character has, for example, as a prior element the lust of power, satisfaction will never curb the need to dominate, whether a group or nature, and this need becomes increasingly strong.

This observation brings the debate back to needs which are individual and which in turn influence the social character and in this way may change the society to which the social character is functional.

Human development, I believe, springs exactly from the dynamics of needs and social character just described and in this sense it can be considered to be the harmonious interaction of needs and social character of the human being in its wholeness in a given society.

In children such human development is crucial in the first five years of life as it determines their future.[4] Looking at human development from the previous point of view, we may see needs being satisfied differently according to the social character of the parents in the family or in the society to which they belong and can, as a consequence, see different levels of development which may not be harmonious and not human. In a society where the social character is based on satisfying biophysical needs first of all, as in the case of the highly industrialized society, the child will grow up healthy physically but with its potentials and needs for love, security, and communication probably underdeveloped (in the case of potentials) and unsatisfied (in the case of needs).

This is of course true unless the social character of the child's specific family will be different from the social character of the society and then this will influence the further development of the child and its sanity.

In fact, although needs like those for happiness, harmony, love and freedom are inherent in the nature of human beings they tend to cause illnesses which may develop into insanity, if frustrated by the social character. When a child's needs for love, communication and warmth come into dynamic relationship with the parents' various social characters, serious contrast may occur. This is the case of the capitalist society where, as Fromm puts it 'each person is a package in which several aspects of his exchange values are blended into his personality'.[5] I interpret this as the package of different social characters which, in the case of capitalist society, may cause abstraction from the global world with 'other social characters'. With such an abstraction and contrast a parent can live, but a child suffers from it. 'Mental health is characterized by the ability to love and to create',[6] and this may be destroyed or at least hindered by the family, local, national or continental social characters of a child living, for example, in Western Europe.

When the basic values of the social character in a given society are centred on the enterprise, when exchange is crucial and action has to be worthwhile, such social character will not satisfy the needs for love and creation and will not foster human development of the whole human being harmoniously between needs and social character which is so important for the child.

As Fromm puts it, in modern man and woman a value which is crucial is anonymous authority related to the law of conformity. The result of anonymous authority and the consequent laws of conformity can be seen in the next part of this chapter where we shall analyze the dynamics of childrens' needs and the social character through their visions of the future.

Although children as persons, in their relation with other people, strive first for security in their roots and then towards development which leads to their choices of life, we shall see that their visions of the future often do not show different choices from those of their parents. This reinforces the discrepancies of needs and social characters centred on the previously indicated 'exchange values'.

In striving towards safety children minimize fears and loss. In striving towards development they maximize attractions and delights. This is in fact an opposite indication to the exchange value which we shall see in children through their visions of the future.

2 Childhood needs and social character expressed in visions of the future

In a previous paper I described the result of research on children's visions of the future. I would like to analyze such visions in terms of needs and social character, that is, in terms as I perceive it, of human development.[7]

The experience of industrial society is related to children of a very poor socio-economic suburb on the outskirts of Rome (Fiumicino) where immigrant unskilled workers from the south of Italy come and live with their families for a couple of years in search of a job that will allow them to enter the industrial society and the big town. They either find a job and leave for Rome or go back to their home village. The children involved in the project were between 9 and 11 years of age and they were asked to draw their answers to the following questions:

(a) 'Me in the Year 2000.' The children describe themselves in front of a fruit cart where the prices are too high for them. Or they describe themselves in a broken-down house; only in one case out of thirty children was a peaceful country scene described. It is an experienced 'me' that is projected. It is a future where the fruit is too expensive to buy and where homes are ugly places to live in.

(b) 'The Room in Which I Eat in the Year 2000.' Here mothers are always working very hard but in some cases they are aided by technological equipment. Again, it is a projection of perceived present.

(c) 'My School in the Year 2000.' The school is later a place where the teacher is very big and high up in the picture and the children very small or the school is a place that is enlarged. The drawings in this case indicated the relation with the institution where children seem to see immutable authority in the teacher seated high up while the children are well below.

From this experience we can see that when a child of six perceives itself in the future of a broken-down home, it shows quite clearly that in the internalized values of the society in which it lives, the home is very important, and it also shows that the need for shelter is not fulfilled: the human development of that child is definitely hampered by the society in which it lives, and does not answer to its needs.

When a child describes its school as a highly authoritarian one in the future, this shows that the social character of the society in which it lives is based on authority values and although the need of the child is for something different (understanding), it does not perceive any other reaction but that of enlarging such an authoritarian school. The human development of the child, complex, integrated at the material and non-material levels, does not seem to be possible because it only wants to destroy the reality in which it lives.

In the cases analyzed, only one out of thirty children was able to describe a peaceful future, indicating that its human development could be linked to a social character of its family, different from the social character which all children had internalized in that specific society. They in fact belonged to the same socio-economic environment and only the indication of different family character could account for the difference.

Another survey also carried out in Italy showed the images of the future which one never wants to see realized, revolving around three themes: work, family and violence.[8]

The main negative images that prevailed most frequently on this theme were unemployment or having a boring or unattractive job. Exactly half of those interviewed identified the above as being the worst or one of the worst images of the future.

The second theme expressed was the absence of a family and/or the impossibility of creating one. Within the internal distribution this fact was more significantly affected by geographical position and class status than the problem of work.

Even though we know the family commands a strong position as a traditional institution in Italian culture, it is surprising to see this same feeling reproduced in images of the young people interviewed.

Therefore, 'not getting married' becomes equally present as, or slightly less than, loneliness and it can be deduced that the underlying anxiety is that in a future world what may be lacking most is love.

The problem of affection and the search for it would modify the significance of marriage certainly in less conformist terms and bring it nearer to what the youngsters interviewed felt.

The fact, however, remains that this institution is still important as shown by the different distribution of this image, which as in the case of work, is a function of income and culture.

More generally and holding the worst future caused by wars, violence appeared in third place for frequency and this seems to be the main anxiety of collective future images.

Recalling just for a minute the latest acts of violence both in this country (Italy) and countries nearby, we can grasp the reality of such anxieties.

In the internal distribution, this fear was noted more widely in Rome, especially in the lower classes. The youngsters living in the capital are not only subjected to the violence described in newspapers but also to the everyday violence peculiar to a large city and this seems to us to explain their tension concerning this problem.

The images of a better future reflect mostly those noted previously: work, family, violence. Even so in this case they present a rather reduced spectrum of contents.

Work holds a great positive value concerning images of the future and this is indicated by the higher frequency with which this point is identified as the image of the best future. In our opinion, this confirms the anxious basis which recurred concerning a negative future and therefore demonstrates just how much the present economic crisis weighs on the replies given. As in the case of work, the family or marriage is also a confirmed desirable image for the future.

Certainly a fairer and non-violent world is wanted by those interviewed. Their social sensitivities are confirmed by references made to peace, right to life, respect for nature as well as the desire for those in the public eye to be more democratic.

The visions of the children (aged 11–13) of the future were linked, as we have seen, to work and stable work and demonstrated how important such activity is in the social character of an industrial society, and how much the children felt their satisfaction of human needs linked to work. This indicates how much human development is linked to work. Stable work first of all is important for survival, for security, for peace in a family and as such is perceived by the children.

The same children as a second element of their vision of the future saw the importance of getting married. Here again the social character linked the answers to their needs, to the institution, but mainly to the security, peace, and tranquillity that marriage seemed to give to the children. Human development hence seems to be linked unconsciously to the life of the family in these children, and when in the questionnaires the children replied that it had to be a marriage of one's choice, human development seems definitely to emanate from freedom and not from compulsion. In this case what appears to emerge is that the social character of the social systems (various, as described in part 1), as transmitted by the parents, is one of compulsion. The further fact that only a few children in the sample dared to think of work they liked, shows how human development may have very different levels in apparently the same social character.

The perception of human development is deeply related to the social system and hence to the social character. We shall see what are the hindrances to human development further on in this chapter.

Children especially, when stimulated either in school or by activities which are outside the social character of their society, show that they may develop as human beings outside and beyond the dictated elements of the social character they perceive in their family and the other social systems. This is the case of an action research carried out in Naples.

Naples, a rather large town in the south of Italy, is part of the industrial society and although it is not an industrial town in itself the fact that it is a port and the largest town in southern Italy involves it in the industrialization process. The children I shall speak about live in Vomero, the high part of Naples, and Rione Traiano which is peripheral. Both are socio-economically deprived areas where people mostly live on part-time jobs, day-to-day work, side jobs related to the port or to selling in the streets.

A project is currently being carried out, also supported by the local administration authorities, and goes along the following lines as aims of the project:

1 To develop positive attitudes of children towards self-direction and research;
2 To expand consciousness of local environment and social realities;
3 To develop children's naturally intuitive, integrative and global ways of doing and perceiving things.

These aims are followed through the use of communication media; the children make their own films and use their own projects, producing their own programmes (slide tapes, audiovisuals, etc.).

Additionally, in their projects children are given the possibility of showing their audiovisual work to the 'sector' (quartière) and in this way they really become part of the environment and have a role in it; that of pointing, in their view, to the problems of their sector. In this sense children become actors of change.

In the case of children living in deprived environments, it shows that alternatives to their social character and to that of their parents are possible. They seem to be able to perceive a different development. The case of Naples, in fact, demonstrated that the children felt the possibility of a richer, fuller development, if stimulated to think of alternatives beyond their social character in that specific society.

Other elements of possible development seemed to emerge from childrens' drawings on the future. I shall try and analyze them.

The experience I want to describe is one carried out by an Italian Association, Eta' Verde (Green Age), which unites children and teachers in the analysis of what have been called 'macro-problems' which are really global problems both in the sense that they involve all the world and are complex problems. This experience involved analysis of the work carried out (discussions, drawings, paintings, sculptures) by children all over Italy with no sample criteria. The children are from 6 to 16 years old and belong to various socio-economic as well as cultural back-grounds. They are related to the school structures since the requests for drawings, etc. are made through the schools. Once a year the children prepare an exhibition of their drawings, paintings, photographs and other means of expression, as well as essays. All are read or analyzed and prizes are given.

The children describe their feelings in what almost by tradition, even if a recent tradition, we call global problems: world hunger, ecological destruction, nuclear holocaust, physical violence, spreading terrorism.

In this case, children of around 6 to 15 years of age, describe their future as dangerous and unhappy. Hunger, atomic bombs, violence, war, etc., are part of their visions. This shows that the social character which the children receive from their parents and teachers is certainly in contrast to their needs which are more related to survival and love. How can we speak of Human Development when children are surrounded by violence and fear? This becomes even more evident when children are prompted to describe alternatives. The alternative to hunger is solidarity (described as children holding hands). The alternative to towns full of pollution is a country scene. The alternative to industries is grass and flowers. The drawings clearly reveal the discrepancy between social character and the satisfaction of needs, and human development does not appear.

The dynamics between the social character and the emergence of the needs has to be analyzed and acted upon for change if we want it to produce human development.

3 Human development, its hindrances and possibilities in childhood at the:

(a) Micro level: the family

In the preceding pages we have described how we can speak of the social character of the family. The family is in fact the agency which transmits to the child the requirements of the society. Parents are the carriers of the social character of the society and, as such, they transmit it. At the same time they are carriers of the social character of the family. On the other hand, the parents transmit their own personal character whether of loyalty, anxiety or happiness to the children.

When we described the vision of the future linked to the children's stable work, we underlined the importance of work in the social character of the society, and also the anxiety of the parents without stable work. The family will be linked both to the social character of the parents as members of a different social system and of that specific family, and to their personal character.

(b) Meso level: group

By meso level I mean the group level and in the case of children, mainly the school.

Here we find the importance of the training methods. But such methods by themselves will not influence the child. What will influence the child's human development is the content which is being transmitted. As an example, a child's description of a school where the teacher sits on a high bench, shows that the method involves authoritarianism as part of the role of the teacher.

This implication is also in the answers of the children to questionnaires, saying that school is only builder of a passage to life and work and not construction of contents, or a transmitter of values. This indicates that the school is linked to a role of maintaining the social character of a society where values and contents are to be kept and in this sense the school is not a builder of human development.[9]

(c) Macro level: society – world

To quote E. Fromm again, he says that the genesis of the social character cannot be understood if it is only related to single causes, but has to be understood related to sociological and ideological factors.

The former are clear in modern industrial society which creates the Western man or woman: the tendency towards profit, possessions and domination. Such factors, though also linked to ideological principles such as attaining security at all cost, are minimizing risk at the cost of others. The principle that each man seeks his own profit and by this makes the happiness of all, is a guiding principle of human beings in modern society. But does this factor foster human development? As can be seen these are both sociological and ideological factors.

Still at the macro level, the industrial society influences the child with the law of the market. Maybe this is changing partially, as A. Toffler puts it, towards a coming together of the producer and the consumer in what he calls Sector A, comprising all the unpaid work done directly by people themselves, their families or the comminities.[10] This of course is a change which would influence childhood greatly and its human development and we can see cases in all countries of such a trend. But can we speak of a world-wide move in this direction? Is it not *yet* mainly the industrial society dominated by the market which has its own laws and rules over man which is changing? Although Toffler does not claim to speak of the whole world, I believe the question is relevant in future terms.

On the other hand, what also influences human development of children is the growth of science and technology and certainly the problems faced by the scientist are not those chosen for the human development of the next generation, the children. We know about adaptation to science and technology, as science and technology have a force of their own to which, in the best of cases, people have to adapt and the same is true for children, which means the future.

Is this therefore contrary to the enlarging of Sector A as described by Toffler?

Just one more observation: the importance of competition in a market-orientated society does not foster human solidarity which children on the contrary, seem to desire, as shown in the research mentioned above.

In conclusion, many are the indications which seem to emanate from the industrial society as not productive of human development, and if such a society is still growing it will need many changes before the crucial aim of societies becomes really human development.

4 Where and how to act: education for human development

I have described the difficulty which human development means for children and the family, in school, in society, in general. What can be done?

All these levels are interconnected: the social character of the parents is linked to society and to different societies and the consciousness of the parents influences the children. First of all, parents should have the consciousness that their way of acting, fostering and aspiring influences the children, in the same way that their happiness or anxiety also transmits itself to the children. Their learning, as the acquisition of consciousness is important in the sense of acting in a different social character from that of the many societies which seem to relate to the industrial society. This would mean changing the society itself.

As to the school, considered as the group where children spend much of their time, it also has to be based on links between school, children and community to develop the children humanly, in their 'real' environment which is not only the one created by the industrial society. The 'real environment' may be able to foster a different social character and, as a consequence, human development.

All such changes and developments are linked however, to the society as a whole. Are we still only in an industrial society which is based on principles of production and not of social usefulness, not of satisfaction with work, not even of human relations? If this is so, we need a revolution of the social character in the family, in the group that may produce one at the macro level, in the society. On the other hand, are there counter trends appearing which will help human development in children, influence the social character of the family and school and foster a new generation of self-actualizing people where human development emerges from a different dynamic of needs and social character?

This is the basic question for human development which has to be faced even if the how is not as clear as the where. I personally believe strongly that the changes towards human development spring from a special relation between needs and social character and have to take place at the family (micro) and school (meso) levels in order to influence the macro level. As this sort of change needs time, it is never too early to begin. We have seen how the macro is difficult to change, but it can be done, as history can show us, and this is the real building of the future.

Notes

1 Fromm, E., *The Sane Society*, London: Routledge & Kegan Paul, 1963, p. 78.
2 Ibid.
3 Maslow, A., *The Farther Reaches of Human Nature*, New York: Viking Press, 1972; Mallmann, C., 'Research and Human Needs Programme', paper prepared for UNESCO, May 1977.
4 Cornelia Quartie, *La Condition Enfantine: ses exigences, ses difficultés*, Rapporto di Congreso, Valencia, April, 1979.
5 Fromm, E., op. cit., p. 69.
6 Fromm, E., op. cit.
7 Masini, E., 'The Role of Childhood in Different Development Styles: the Future Seen in the Present', paper prepared for UNESCO Latin American Human Needs Meeting, Tiradentes, 1979.
8 Ibid.
9 Botkin, J., Malitza, M., and Elmandjra, M., *No Limits to Learning*, Oxford: Pergamon Press, 1979.
10 Toffler, A., *The Third Wave*, New York: William Morrow, 1980.

10
Human Development and Popular Culture in Latin America – Case Studies

Maria Teresa Sirvent

Introduction

The data used in this chapter is the result of the studies undertaken by the writer in two communities, one located in suburban Buenos Aires (Argentina) and the other on the urban outskirts of São Paulo (Brazil).[1]

The population studied in both experiments belongs to the social categories of urban or suburban areas commonly known as popular sectors: workers and 'marginal' groups, including the lower levels of the middle sector.

The composition of these popular sectors, however, is not similar. The urban peripheral community of São Paulo is almost entirely made up of marginal or peripheral social groups which endure harsh situations of social and economic need. The case of the Buenos Aires suburban community, on the other hand, shows a larger proportion of a popular sector comprising workers and lower middle class levels with a minimum degree of satisfaction of needs such as housing work, health and education.

In the following pages an attempt will be made to present some aspects of the problem of poverty in Latin America in relation to a desirable model of human development. Attention will be focused on those sectors of population which inhabit Latin American cities in conditions of poverty and even extreme social and economic poverty.

Generally speaking, the urban or suburban areas of Latin America display in varying proportions the dramatic reality of belts of poverty or social periphery.[2] These areas constantly receive enormous streams of internal migration composed of rural population groups who regard the city as the solution to their deficient economic or social living standards. These migratory groups are those which mostly form the great human core of poverty belts in Latin America. In São Paulo, for example, these belts comprise 75% of the total population.[3]

We find ourselves face to face with strong contrasts, like the two sides of a coin, between growth and poverty, luxury and pauperism, overconsumption of superfluities and underconsumption of essentials. The most poverty-stricken popular sectors are conglomerates of low-income population with serious problems of unemployment and labour instability, low standards of education and minimum or non-existent professional qualifications. These populations with low income, prestige and power, settle in or are continually forced towards zones far from the well-equipped centres, so that they occupy the most rundown areas of the city, generally lacking minimum services of health, transport, communications, water, light, roads, etc. This kind of situation is aggravated by problems of cultural shock, the lack of popular organizations for change through collective action, and scant recognition of the need for participation and group work to overcome the problems of everyday life.

These reflections arise from three basic questions:

1 Compared with a desirable model of human development, what are the features of the forms of daily life of the popular sectors in Latin America? What is the distance between the desirable parameters and the daily routine?
2 What aspects of this daily routine can operate as factors that facilitate or inhibit the human development of the most destitute groups of Latin America?
3 In the present situation of Latin America, is a cultural action feasible and viable which aims at overcoming or minimizing those aspects of everyday life in the popular sectors that act as mechanisms inhibiting human development?

The idea of human development adopted here basically implies raising the quality of life of the population and, particularly, the poorest popular sectors of Latin America. The concept, quality of life, differs from concepts such as level or standard of living to the extent that it connotes an integral reference to the whole set of human needs and not just part of them. It refers to the equitable, fair distribution not only of satisfiers related to basic, obvious needs – such as health, housing, work, food – but also those resources which society possesses at a given historical moment for the satisfaction of non-material or not so obvious needs, such as playing the leading role in one's own life story.

The way of satisfying one need influences the other needs. This means that if to satisfy the housing need dwellings are provided in the form of assistance, we are inhibiting the satisfaction of such needs as participation or reflection on the facts of everyday life. It does not mean abstaining from confronting the solution of a basic need like housing or work, but rather it implies tackling the solution of the problem through participative strategies which gradually develop reflexive thinking and community creation.

The criteria for determining non-material needs which contribute to the population's quality of life arise not only from considerations based on human sciences but also on value judgements referring to the image of a person or a desirable society. The human need to participate, i.e. to have a responsible role in one's own life, is in this sense one of the non-material needs recognized as a condition of life of a population. Real participation of the population in decisions which affect their daily life also assumes the recognition of other related needs which, in turn, are a condition resulting from a participatory process: self-appraisal and appraisal of the culture of the group to which they belong, ability to reflect on the facts of everyday life, ability to create and re-create new forms of life and social co-existence.

In the socially peripheral groups there may be a gap between the existence of real needs and their recognition by the population, especially as concerns non-material or not too obvious needs. We are referring to the so-called 'social trap' according to which, to the extent that the degree of social and economic impoverishment increases, recognition of the real needs diminishes. Thus, for example, the impact of the consumer society on the groups which make up the Latin America urban periphery can produce in them a series of 'false needs', generating appetites that push the groups towards behaviour patterns which do not favour solution of the daily problems but, on the contrary, often aggravate them. In the living patterns of the popular sectors, representations or images of the facts of daily life can be encountered which hinder recognition of real needs such as participation in the decisions affecting day-to-day living.

All social, cultural or educational action leading to human development must start out from the differences between *subjective and objective needs* and must be aimed at modifying the inhibitory representations and images of such recognition. Subjective needs refer to the perception of wants by individuals or groups of individuals. Objective needs point to wants of individuals and groups which can be determined *independently* of whether the individuals concerned are aware of them. Both types of needs may or may not coincide. If they do not, the gap referred to above appears.

Conceptual differentiation between objective and subjective needs is important both theoretically and for its implications in the determination of objectives for a transforming action. An educational or cultural action which rests only on the manifest needs of a community runs the risk of putting into effect activities which do not lead to transformation of the existing situations of imbalance.

To summarize, from a social angle human development is understood here as a process which favours minimization of the acute imbalance between an economic, political and cultural élite responsible for taking societal decisions, and the majority of the population which does not participate in decisions affecting its everyday life. From an individual angle, this process demands the growth of the men and women of the majority popular sector who would 'feel' their non-material objective needs and, in particular, the need for participation; who would increase their ability to reflect on the causes and consequences of their daily problems; who would appraise themselves and their own groups as able to bring collective and potential forces for change; who would have the possibility of creating and re-creating new forms of co-existence that would overcome the sharp social inequities of present-day Latin America.

The contrast between this desirable model of human development and the daily life of the popular sectors studied in Latin America is profound.

We find:

(a) An important sector of this population without real participation in the transformation of their world and in overcoming the conditions that affect daily individual or group life.

(b) The existence in these sectors of conceptions or images of social participation that are obstacles to the emergence of groups which would be protagonists of their own history. For example, the vision of participation as something to be feared or rejected, or the internalized acceptance of authoritarian or paternalist models as the only possible forms of social relationships.

(c) The presence of organizations or associations of popular participation which are a reflection of these conceptions and therefore reproduce within them the structure of social inequality of the global society. For example, neighbourhood associations which operate according to an élitist, authoritarian model of decision-making.

(d) The shared presence of images, concepts, meanings, value systems, which may inhibit the possibility of reflective thinking on the facts of daily life and hinder the emergence of actions for social change. This is the case of a magical, determinist conception of life which excludes the possibility of a reflexive, creative human action for overcoming everyday problems. In other words, conceptions or social images which attribute the cause of daily problems to 'fate' or to the 'inferior nature' of the destitute populations, and therefore justify any feeling of individual or collective impotence.

(e) A minimum recognition of needs of participation, collective self-appraisal, reflective thinking, creation or re-creation as being fundamental needs, whose satisfaction is an essential condition for a process of raising the quality of life of the Latin American popular sectors.

In the everyday forms of life of the popular sectors, however, these inhibitory aspects of a social transformation process co-exist with mechanisms that facilitate and stimulate.

As previously pointed out, this chapter aims at analyzing some aspects of the culture lived by the most impoverished urban sectors.[4] In other words, it refers to the contents of popular culture. Popular culture is understood here in terms of an internalized culture which is reflected in the living patterns or behaviour modes characteristic of the urban popular sectors in the different spheres of the daily task: working area, family, education and day-to-day social life.[5]

In view of these objectives and the complexity of the cultural scene, emphasis has been placed on the study of the various behaviour forms characteristic of daily life in non-working time or time free from compulsory occupation. This sample does not mean that free-time behaviour forms are assigned their own dynamics unconnected with the working scene. On the contrary, the forms and meanings that human behaviour assumes in non-working time are largely explained by the type of reality the person faces in his working life. There are three reasons for the emphasis given here: firstly, starting out from the assumption that it is in the area of free time where the traits and features of a living culture show up with the greatest clarity, and secondly, free time is considered an area of greater social and political feasibility for cultural-educational action in the present conditions of Latin America.

For the analysis of the cultural components of these behaviour forms, three types of variables have been chosen on the basis of the relevant problems from the viewpoint of human development and of the intervening factors on which they can exert a modifying action: cultural practices, shared needs and social representations.

In the analysis of these three types of variables, the following are explored:

(a) The distance between the desirable model of quality of life centred on the emergence of participating groups, creators and protagonists of their activities and the reality of everyday life of the popular sectors.
(b) The joint presence of factors which facilitate and inhibit the participatory, reflective, creative and autonomous growth of these sectors.
(c) The feasibility of cultural or educational action directed at the human development of the popular sectors and in the present social, political and economic conditions of Latin America.

Human development and cultural practices

Ordinary, habitual or widespread activities in the popular sectors during non-working time are known as *Popular cultural practices*.

They cover the whole range of daily behaviour forms, from activities such as exposure to the mass media, the practice of and/or attendance at sporting activities, social interaction with families and friends up to social participation or community association for collective solution of neighbourhood problems. For the purposes of this chapter, the intention is to analyze the presence of spontaneous components of participation, reflection or creation in the cultural practices of the popular sectors. In this sense, our reflections are not centred on the activity in itself, but rather on the predominant type or style of relation of the individual to the objects of the world about him. In this way a distinction is made between a *consumerist style*, typical of exposition to the mass media and which is characterized by the action of passive reception as a spectator, without modifying the objects or goods to which he or she is exposed,[6] and a *productive-creative style*, characterized by a transformer action on the objects through doing, creation, discovery or expression. The collective cultural practices of community association can be an example of the non-consumerist production styles in the daily life forms of the most impoverished sectors. Social participation activities help develop invention or creation in terms of solutions for community problems. The neighbour who participates is reflecting on and looking for new solutions to old problems, new forms of action.

In the area of daily time free from production activity, the relationship we call consumerist is reflected especially through exposure to the mass media. This type of relationship is not peculiar to the media. The consumerist link can occur in any situation of daily life in the family, school and working areas as well as political areas.

Its most outstanding features are:
- The distance existing between the act of creation of the cultural object and its reception; in other words, the subject receives information, images, without taking part in their production or diffusion.
- Unilateralness of the communication process, since the functions of emission and reception are not interchangeable and frequently there tends to be a space-time gap between emitter and receiver.
- Lack of stimulation for reflective thinking, since it is the emitter who poses the problems, analyzes them and maybe solves them. Consequently, no opportunity is provided for training to pose, describe and search out the causes and consequences of the problems.
- The difficulty, as a result of the above features, of transforming the

messages received into action on the individual or collective reality, aimed at overcoming problems.

Contrariwise, a productive-creative relationship is characterized by:

- Direct participation in the creation of the cultural object, whether a material or non-material object, such as a regulation or a value.
- Lack of rigid distinctions between emitters and receivers in the communication process.
- Emergence of reflective thinking, to the extent that the subject acts individually or collectively to solve problems which may arise.
- Possibility of individual or collective action tending to modify the real situation.

Consumerist cultural practices

In the two communities studied, a clear predominance appears of strongly consumerist practices, especially exposure to television and attending sports events, in other words, activities where the principal feature is dependence on an emitter and an external organizer.[7] This situation reveals the distance existing between these aspects of daily life in the popular sectors studied and a model of participative, creative, human development.

In the case of São Paulo, this situation becomes even more acute in connection with women. Analysis of the data shows important differences between men and women as regards characteristic forms of free-time behaviour. These differences reproduce the situation of feminine social inequality in all areas of daily occupation, mainly in the working area. The migrant woman has less probability of labour mobility than the man. She begins and remains all her life in jobs without training and with low wages (domestic service, cleaning streets and institutions, garbage collecting, informal labour market, etc.).[8] The working period for women of the urban periphery in the case of São Paulo is a complex and difficult time.[9] More than 50% of the working women interviewed expressed their dislike of working outside the home and therefore their desire to remain in their own house and neighbourhood the whole time.

The man generally works outside the community, spending at times from one to three hours a day travelling to and from work. For the man, the community becomes a 'dormitory neighbourhood'. In spite of his impoverished situation as regards the working world, he has at least the opportunity of being in contact with informal learning sources, such as the job, the factory, the street, etc. In other words, his daily life provides him with a series of stimulants which favour his adult socialization in the urban world. The man learns day by day a series of expected behaviours which enable him to interact in the 'big city'. Basically, he gets to know the 'laws' of urban life, its institutional regulations, secondary relationships, everyday bureaucracy. The woman, however, remains shut in between the limits of domestic services and her immediate surroundings.

These differences between men and women during their occupied time are reproduced during the free time of the population. In a situation of poverty, in terms of quality of life, the man has the possibility of a greater range of stimulants. A typical Sunday for the woman of the community studied is characterized by the following activities: to remain at home, do household jobs, watch television, look after the children, go to church or visit relatives. In general, for the woman, there are no significant differences between a Sunday or a weekday. Conversely, for the man, he has more opportunity to escape from the daily routine by going to the bar, playing billiards, watching or playing football in the neighbourhood, gossiping with friends in the bar, drinking, attending football matches in other neighbourhoods further away, going to the city centre, etc.[10]

Activity shared by the whole family group is minimum; in addition, the man spends a minimum amount of time with his children.

This reference to daily life of this population reveals the accentuation of social inequalities in the case of the woman. Even a qualitative analysis of activities during free time shows a greater presence of a woman in consumer activities.

Productive-creative cultural practices

In both the communities studied, the undertaking of productive cultural practices is scant. However, although they may not be quantitatively important, the presence in the population of some activities where the productive-creative style predominates, points to the existence of elements which facilitate or stimulate collective activity for change. This is the case of social creation by way of community participation which, although not practiced by the majority, appears as a potential source for overcoming social imbalances.[11]

A case common to both studies is the channelling of social participation through voluntary neighbourhood associations, i.e. associations originated by the residents themselves around mutual interests for solving neighbourhood problems (development associations, friends-of-the-neighbourhood societies), carrying out sporting and/or social activities (clubs), association around shared tasks (Mothers' Unions), etc. In spite of the essentially productive-creative nature of this collective social practice, meeting of neighbours, collective reflection and discussion of problems, community decision-taking, creation of solutions or alternative action, etc., the study of their features reveals the co-existence with them of contradictory traits, facilitators and inhibitors of human development.

For this analysis a reference framework is used, formed by the concepts of real and symbolic participation.[12] It is considered that the members of an institution exercise a real form of participation when, through their actions, they influence the processes of institutional life, for instance:

(a) in decision-taking at different levels, both in the general policy of the association and in the determination of objectives, strategies and specific action alternatives;
(b) in putting the decisions into effect;
(c) in the permanent evaluation of institutional operation.

Participation is symbolic when the population, through its actions exerts only a minimum influence at most at the level of policies and institutional operation. Symbolic forms of participation may generate the illusion of exercising a non-existent power.

In general terms, the neighbourhood associations or groups analyzed are not characterized by the real participation of the population in their operation. Quantitatively, only a minimum part of the population is connected with these associations. They reproduce in the social periphery the inequalities of social participation of global society: there is a small number committed to the association and the great majority is peripheral.

In the case of the Buenos Aires community, both the data of a population poll and the interviews held with association directors, reveal that neighbourhood participation is restricted and when it exists it tends to be limited to issuing a vote in the elections, sporadic attendance at assemblies in which sometimes the legal quorum is not obtained, or else to the 'donation' of personal work for implementing decisions taken by a small group (held in construction or maintenance jobs, collaboration in preparing a fête, etc.). Participation is generally symbolic, as illustrated by expressions such as the following, taken from interviews held with the leaders of some associations.

> Very urgent decisions I take myself . . . sometimes with three others. In general the leaders 'think' and the 'rank and file' carries out the manual jobs.

This symbolic participation can also be found in the participation structures which the peripheral schools of São Paulo have generated for the voluntary participation of the parents. This is limited to attending meetings when they are invited. These meetings are usually organized according to a model which in fact creates difficulties for the real participation of the parents. Another form of participation accepted by the school is monetary collaboration or 'donating work' to undertake tasks decided by the school authority without the intervention of the parents in decision-making. When there is an attempt to invite the parents to 'participate' in a decision, it is the school authorities who present the ideas and put them to the vote, usually without the parents making any attempt to argue with the proposals.

The structures of the voluntary associations – including those promoted by the school – show a series of traits which operate as inhibiting factors for increasing the real participation of the population. These factors reveal how the dominant forms and mechanisms of non-participation or symbolic participation are reproduced right within a popular association. We found:

(a) Presidents who had occupied the position for several years.
(b) A committee which had also been installed several years with only a few partial renewals.
(c) A high degree of bureaucracy imposed by the operation of the institution (formal resources, minutes of meetings, voting, etc.) which hinders spontaneous, informal participation by the residents of the neighbourhood.
(d) Monopolizing decision-taking in the president or a single person probably with the symbolic participation of the rest of the Committee. The process of decision-making is generally authoritarian, with a monopoly of the power, in spite of the 'democratic' formulas.

To have power means having resources. The resources are time, information and knowledge of the state bureaucracies (knowledge of the functioning of some public offices or direct knowledge of some public employees). The member of the community who is able to monopolize these resources generally becomes the irreplaceable 'leader':

> Mr X has the papers, he doesn't allow anyone to read the statutes. (Interview with an active member of the Friends of the Neighbourhood Society.)[13]

The possibility of information, or knowing how to get connected with the public authority, is an acquired behaviour which is not transferred but unfortunately becomes monopolized.

The vertical and hierarchical structure is the socially accepted form of institutional action. The voluntary associations assume it as the only one possible, both in outside relations and their inner operation. Their relations with the public authority only concern 'asking' and 'waiting' for a solution to come from the social, political or economic authority.

All this gradually generates a series of factors which, through the institution, hinder the real, majority participation of the population. Nevertheless, this phenomenon of organizing voluntary associations in the popular sectors presents at the same time dynamic transformer features:

(a) The emergence of groups which disagree with the functioning described above and try to change it. These differences occur fundamentally in terms of disagreements with the monopoly of power and relations with the political power.

Unfortunately, these institutions cannot by themselves bring about the changes and finally the stimulator groups, if they assume power, also assume the characteristics of the conservative groups in terms of authoritarian handling and monopoly of decisions.

(b) The existence of educational potential, both for the leaders and the more active members. This potential is expressed fundamentally in spontaneous learning about relations with the public authority, institutional 'laws' and apprenticeship for participation, communication and expression of ideas.

(c) The participation of women is another of the outstanding dynamic aspects. In spite of being the object of discrimination, the woman, especially in the Brazilian study, assumes leadership of serious, deserving neighbourhood movements (for crèches, schools, infrastructure, etc.). The case of São Paulo is an example of the new role which the woman of the urban periphery is beginning to play in the social struggles to improve living conditions.

(d) The existence of certain experiencing of self-appraisal and/or appraisal of the neighbourhood forces as the result of collective actions.

Unfortunately there are no mechanisms for the transmission of these apprenticeships which, on the contrary, are monopolized, thus inhibiting their transformation into a generalized experience of capacity and autonomy for change.

All cultural or educational action should unfold the dynamic aspects of this productive-creative and collective use of free time, modifying its inhibitory aspects reflectively.

Intermediate cultural practices

If the productive-consumerist 'axis' is taken as a continuum, intermediate forms can be found which have only some of the features of the extreme ends. In other words, the way from mass consumption of television to social participation passes through zones of intermediate activities such as the practice of sports, open-air outings or visits to friends and relatives.

In both the communities studied, these intermediate activities are widely carried out, oscillating between one end and the other of the consumption-production axis and on general lines they are characterized by 'doing' more than by 'watching'.

For instance, there are outings, social interaction activities (within the family or outside), social games (cards, billiards or draughts), practice of sports and handicrafts where imitation is more usual than creation or re-creation. The importance of these intermediate practices lies in the potential presence of a series of participatory and creative connotations which can be developed and enriched through an action programme aimed at human development.

In the working class zones and the shanty-town sectors of Buenos Aires, the intermediate practices which predominate are social visits to relatives, social games – dice, billiards, bowling, quoits, table football – young people and neighbours gossiping on street corners or in their houses, dances, folklore groups, playing football and boxing. Several of these cultural practices bring together rural cultural tradition handed down from one generation to another. 'More urban' games or sports – chess, draughts, table tennis, basket-ball – only appear in the working class groups from earlier migrations, more stable and having skilled or semi-skilled labour.

The qualitative information of the São Paulo community shows a greater presence of intermediate activities by the men, such as the bar, billiards, draughts and football. In the case of the women, the predominant intermediate activities refer to family social life – visits to relatives – or carrying out feminine handicrafts such as embroidery, knitting or dressmaking which, in spite of generating a material product, may show different degrees of creativity.

A series of common traits in these cultural practices of the popular sectors is evident:

(a) In general terms they are carried out within the geographical limits of the neighbourhood. While it is considered that this characteristic of physical proximity is common to all the popular sectors, it is accentuated in the peripheral neighbourhoods due to communication difficulties.
(b) Primary type, face-to-face relationships prevail.
(c) They are activities which require the participants to undertake almost no prior planning, preparation or projects which extend into the future, with the exception of football or billiards championships. They are also distinguished by their low degree of structuring as regards the presence of a formal coordinator, a programme, or extremely strict timetables or rigid control rules.
(d) The majority of these cultural practices of the peripheral sectors are in the nature of games: the intention is to seek entertainment and while the time away.

These traits of the intermediate cultural practices are associated with general characteristics of popular sector modes of collective life, such as physical or temporal immediacy, the prime importance of the present, of affections and the rejection of formalized relationships. The primacy of close space corresponds to the primacy of the present. In the same way that the projects of the popular sectors are short-term projects, their use of space is intense and reduced to the neighbourhood boundaries. This restrictive use of space is related to the mechanisms of solidarity and primary affection which unite the inhabitants of the neighbourhood. Solidarity, a characteristic proper to social relations in the rural community, is prolonged into the peripheral neighbourhoods of the urban popular sectors.

In the case of São Paulo, the structures of solidarity and communication proper to rural communities have been clearly preserved, unlike other Latin American cities in which they have often deteriorated to the extent that the rural migrants become integrated into urban life.[14]

Solidarity is understood in terms of a mutual giving and receiving in moments of need and is expressed through informal relationships of direct help to the neighbour or affective and material support in the face of tragic personal situations (illness, death, fire).

> . . .here the people try to help their neighbours. I always say: my closest relative is my neighbour. . . (interview in São Paulo)

This solidarity ensures an informal grapevine among the residents which allows a constant circuit of information. However, it does not become a formalized solidarity in terms of community participatory and organizational structures for solving problems. For this reason, several residents indicated a greater union or friendship in the face of common problems as being among the needs of the community.

In the case of Buenos Aires, the qualitative information shows that informal solidarity sometimes becomes really productive on passing from a temporary, individual resolution to the organization of sporadic collective actions for the solution of a community problem.

Nevertheless, the qualitative data permits risking the hypothesis that this informally organized solidarity is choked off to the extent that the formalization of association structures progresses. Probably this phenomenon can be associated on the one hand with the rejection – common to the popular sectors – of very formalized relations and, on the other, with the bureaucratic and authoritarian forms and the monopoly of power which the association structures gradually assume and which hinder the possibility of development or extension of basic solidarity to more ample areas of participation.

The intermediate practices, potential sources of production and community creation which are characterized by physical and affective proximity and the rejection of very formalized relationships, reveal traits of popular culture which should be taken into account by all cultural-educational action aimed at minimizing social inequalities.

To summarize, analysis of the cultural practices in the communities studied shows:

(a) The prevalence of a consumerist style which is reflected in cultural practices characterized by the distance existing between the act of creation of the object and the act of its reception, the unilateralness of the communication process, the lack of stimulation for reflective thinking and the consequent difficulty of transforming the messages received into an individual and/or social action aimed at overcoming or solving problems. This situation shows the gap between reality and the desirable model of participative and creative human development.

(b) The scant incidence of a productive-creative style. However, social creation by way of community participation, although numerically unimportant, presents a series of traits which show it to be a potential source of a more significant production and community creation. These traits are the emergence of critical groups, social apprenticeship of the leaders as mediators between the neighbourhood and global society, participation of women. A series of characteristics appears, however, which hinders the real participation of the population, reproducing a receptive and assistential 'consumerist style': symbolic forms of participation; existence of a minority which decides and a majority which implements; rigid, bureaucratic structures; authoritarian processes of decision-making with a monopoly of power.

(c) The importance of intermediate cultural practices, characterized by 'doing' rather than 'watching' and which range from one extreme to the other of the consumption-production axis. The most outstanding features of these practices are: they are generally held within the geographical limits of the neighbourhood; a predominance of primary face-to-face relationships; a low degree of planning and their game-playing character.

(d) The presence of a basic ingredient of popular culture: informal solidarity which is expressed in everyday help to the neighbour and sporadic organization of collective actions (claims on the public authority or entertainment in the neighbourhood). This generalized feature of the communities studied is one of the factors which could facilitate the emergence of groups having responsibility for the daily activities.

(e) The accentuation of situations of social inequality in the daily life of the women.

Human development, needs and social representations

Collective needs associated with cultural practices

Analysis of the practices of the populations studied from the perspective of the predominant type of subject-object relationship, reveals the lack of participation and collective creation activities in the face of day-to-day problems. This implies the existence of unsatisfied objective needs, independent of the fact that these needs may be subjectively felt as such by the people concerned.

If it is assumed that the quality of life and collective human development are associated with a balanced satisfaction of the non-material needs of participation, reflection, creation and re-creation, it is necessary to explore the distance existing between the needs expressed or recognized by the population and the objective needs of participation, reflection, creation and re-creation.

The data from both communities shows a low incidence of the subjective needs of participation, reflection, creation and re-creation; in other words, scant recognition of these objective needs on the part of the popular sectors analyzed. The manifest gratifications obtained by individuals or groups in their different activities have been taken as the expression of a subjective need, i.e. what is being sought when doing this or that activity?

Information from the interviews both in São Paulo and Buenos Aires reveals that in the predominant consumption activities there is a search for evasion, escape, for immediate emotions – romanticism, tenderness, suspense, action. This primacy of the search for immediate emotions in the consumerist cultural practices of urban periphery sectors does not constitute an isolated feature but rather it is associated with the popular sectors' perception of time, linked to affective immediacy and the impossibility of postponing compensations. The contents of fiction, characteristic of a search for immediate emotions, are generally accompanied by a high degree of affective surrender through identification with one or more characters – the hero, the heroine – and the establishment of an imaginary interaction with them.

The search for satisfaction of the needs of participation, reflection about reality, creation or recreation is not only hardly mentioned but is also concentrated on cultural practices rarely undertaken by the population. This is the case in São Paulo of the gratifications sought by some active members of community associations and which refer to apprenticeship and personal enrichment.

There is pleasure and gratification arising from participation itself. Although this only reaches a minority of the population, it reveals the presence of a certain recognition of non-material needs which contribute to greater quality of life.

The predominance of the subjective need for escape or evasion may be considered one of the indicators of the so-called 'social trap' or 'tragedy of the common man'. A series of situations and stimulants occurs in urban life which prevents recognition of the objective needs for participating, reflecting, creating and re-creating collectively. There are, nevertheless, as we have said above, indicators of a recognition, however, small, of these objective needs. In the case of the community studied in São Paulo, these indicators appeared over and above the cultural practices when perception of the community in relation to community needs felt as having priority was explored. Although the population expressed consensus in pointing out survival or subsistence needs – security, material infrastructure of the neighbourhood, health – explicit recognition of the needs associated with adult education, and re-creation or use of free time by adults and young people, appeared in second and third place. Moreover, several mentions are made, although in insignificant numbers, of the need for participation, discussion or collective reflection, for group action for overcoming daily problems.

Social representations connected with cultural practices

The choice of cultural practices as satisfiers of collective needs does not respond only to a given needs system but also depends on other factors. The predominance of a consumerist style, the accentuation of inequalities of women, the contradictions in the very interior of the participation structures of the community, minimum recognition of the needs of participation, reflection, creation and re-creation are associated in the popular sectors with the presence of representations or visions of the world that inhibit the emergence of protagonist groups committed to their historical evolution. *Social representation* is understood to be the set of concepts, perceptions, meanings and attitudes which the individuals of a social group hold and share in relation to themselves and to the phenomena of the world around them.[15] The importance of the psycho-social idea of representation is based on the fact that it points to a socially-shared vision of the surrounding reality. It is not a momentary or fragmentary opinion but rather the structuring of a wide range of information, perceptions, images, beliefs and attitudes prevailing in a given social system.

These social representations, understood as organized universes of opinions, beliefs, attitudes, knowledge, are socially determined by the economic, political and social conditions which affect individuals and groups in different ways. The representations elaborated by Latin American popular sectors are socially determined centrally by the conditions of social and economic poverty and by the relationships of subordination to the dominant classes or groups. They are rooted in the structural conditions of daily life and contribute to strengthening them.

Social representations, while at the same time allowing individuals to find their bearings in the world, may inhibit the preparation of social transformation projects. For example, social representations of women, men or the family, reproduce or reflect the experiences of authoritarianism and subordination in which they are structured, thus inhibiting a widened social participation of men and women in the facts of their everyday life. This is the case of the social representation of women detected in the São Paulo community. The feminine image is inhibitive of the recognition of the needs of participation, reflection, creation, self-appraisal and any other behaviour capable of modifying a life of feminine domestic routine. The woman is perceived only as a woman in her home, busy with the domestic chores, her children and her husband. The man clearly displays his pride in the maintenance of the home and his fear at the possible feminine independence.

Both men and women accept feminine work only when the woman is single. To marry is synonymous with giving up work. The majority of the woman interviewed worked when they were single but gave it up upon marriage.

In addition, while the need for feminine work can be accepted in view of economic exigencies, the man tends to fear institutionalized work in factories, thereby boosting domestic service or infra-human tasks as the only alternatives.

The woman expresses in general terms her acceptance of this situation. She even makes veiled criticisms of the women who work. There are exceptions, which are not casual, in the women who participate in feminine groups such as Mothers' Unions or the 'neighbourhood unions' of women. These are the only ones which express another vision of women. But the majority show acceptance and resignation. This image has not only been transmitted but is also learned. A woman's level of aspirations as regards change and work is low.

When a woman accepts going out to work, she does so only for economic reasons. Exceptionally, she also mentions the need to break out of the daily routine. And when she refers to looking for a job, she is thinking only of domestic service. Her 'job', her speciality, from time immemorial has always been in household tasks, and she does not see herself as having potential abilities for any other type of employment. These representations carry with them a strong component of devaluation of the feminine image.

Social representation of the woman is strengthened by the way in which the woman of this São Paulo community regards 'her' man. The feminine vision of the man is negative. There exists a general consensus about masculine 'indolence'. Therefore the woman assumes that if she works the man will take advantage of the opportunity to work less or not at all.

Thus a series of perceptions, visions and beliefs appear which only 'justify' the forms of daily life. The devaluation component which frames the image or social representation of the woman is clearly apparent when the man refers to social participation of the woman or of more active women. In all the expressions there is a connotation of rejection of social activity by women.

Nor does the woman have a positive image of the family or of family relationships. The majority of the expressions imply some sadness, pain and suffering. She suffers in silence, resigned to the feminine condition. In addition, alcoholism, desertion and rejection of the separated woman, are real situations which contribute to the negative images of woman and family.

The predominance of a consumerist style in the whole population, the minimum recognition of objective needs of participation, reflection, creation and re-creation are also associated with the social representations of the causes and solutions of the problems of the poverty-stricken groups, social representation of participation and social representation of culture or of the creative act. These representations act as inhibitory mechanisms or forces for recognizing the non-material objective needs and for the growth of the productive-creative style of collection action. A social representation is inhibitory to the extent that it hinders recognition of objective needs and of all reflective processes or awareness. A fatalistic representation of life, for instance, does not allow facts to be regarded as modifiable, and above all, modifiable because of individual or group action. Nor does it facilitate recognition of participation as an inherent need of the human condition.

Analysis of the causes to which problems are attributed reflects a primacy of magical causes – fate, instinct, birth, 'life's like that' – and of blame laid on the individual himself or on characteristics of the group to which he belongs. This presence in the majority of a magic, determinist attitude, inhibits the possibility of recognizing needs of participation, reflection, appraisal and creation. Thus there are attitudes of conformism and acceptance of situations. Parents, for example, tend to attribute truancy and grade repetition to their children's problems of incapacity. The problem of truancy and repetition is seen as something 'natural' and school success as something extraordinary. This inhibits the possibility of reflection and thus of participation in the solution of problems.

This perception of causes is associated with a mention of solutions centred on authority, i.e. with an assistentialist view of problem solving. There is little mention of the possibility of solutions originating in the community. Everything must come from the competent authority, or the religious authority and therefore the claim is made to authority, or else it is just hoped that the authority will solve the problem some day.

This assistentialist vision of solutions and of the public authority is reflected in the structure of the community participation associations and is connected with a social representation of participation and its forms. In the work carried out in Buenos Aires, the opinions given by the majority of the directors of local associations refer to a series of aspects which are seen to be clearly related to each other and which allow risking the hypothesis that the leaders generally possess a representation of participation which coincides with the above-mentioned characteristics of rigidity, authoritarianism and bureaucratization of institutional operation. Consequently, this representation may be one of the inhibitory factors both of the incorporation of new participants and the transformation of participation into a form of real participation. When the leaders refer to the participation of the community, they are thinking of forms which make it obvious that they do not perceive any differences between the meaning of real and symbolic participation. Save in exceptional cases, it cannot be seen from their comments that they hope their neighbours will co-participate with the members of the governing committee in the decisions, planning and evaluation of institutional life. The most generalized expectation is connected with symbolic forms of participation, carrrying out jobs decided by others, making use of the installations or services of the institution and contributing with money.

In the case of the São Paulo community this vision appears again. To participate means, for the leaders of the Friends of the Neighbourhood Society, that the rank and file put into effect the decisions previously taken, without their participation, by the leaders.

The opinions of the leaders about the reasons for non-participation in the institution reflect, in general, a negative image of the personal characteristics of the residents. Their opinion is that the lack of participation is the effect of indifference or lack of solidarity. This type of appraisal, which explains the low level of participation in terms of subjective uninterest, without taking into account factors external to the person, is the most prevalent in both the communities studied.

The voluntary vision which attributes non-participation to the people's 'lack of willingness or interest' and participation to the fact of 'having been born with a fighting spirit', inhibits reflection on the structural and institutional causes of the lack of participation and hinders creation of new forms which would increase it. It is only conceivable that 'the people' should go to the institution and not vice versa. Thus, both in the case of São Paulo and Buenos Aires there are very few leaders who attempt to understand the structural reasons for the slight participation.

A few leaders of the Buenos Aires community pointed out that the lack of participation is generally associated with a depreciatory perception of the possibilities of changing reality through direct participation in community affairs; with an attitude of fear to the risks participation may involve; and with a negative image of the leaders. In the case of São Paulo the 'non-voluntarist' reasons outlined only refer to the population's unfavourable or unattractive image of the institution due to the doubtful handling of funds or to clashes between some members of the population and the authoritarianism of the leaders.

In short, in the two communities, both the leader's representation of participation and that of the population fail to contribute to collective recognition of the need for participation or the questioning of their present forms. Nor do they facilitate the generation of new structures of community association.

Another of the social representations which may inhibit the emergence of creative and re-creative groups is that associated with creation, education and apprenticeship. In the Buenos Aires case, these ideas all refer to an academic representation of culture which is inhibitive of the appraisal of popular creations and apprenticeships. There are two types of representations of culture in the populations studied. The most widespread is that which assigns it a meaning of an academic nature. It is associated with books, conferences, encyclopaedic knowledge, highly specialized knowledge, higher level formal education, mastery of the language as a form of communication, intellectual and abstract handling of ideas, artistic sensitivity, etc. Only a minority of the population assigns to culture meanings more closely associated with daily life. Academic representation of culture refers to 'packages' whether of information or behaviour rules, accumulated and transmittable, with a universal value, which the subjects must absorb or assimilate in order to become cultured people. Hence this academic image generally materializes in cultural activities or practices which are much more notable for their consumerist rather than their productive components.

Academic representation also tends to be associated with innate conditions for acquiring culture, with the importance of the school and books as means of cultural acquisition and with a clear division between cultural and non-cultural activities, between entertainment and culture.

The representation least connected with academic culture gives a less universal vision of culture, more closely related to the different social groups. It is in this image of living culture where the idea of culture as a creative, transforming attitude appears, contrary to the mere consumption or reception of 'packages' of information external to the person. Scholastic institutions or books are not any more the fundamental means of cultural acquisition. All lived experience is seen as a source of cultural apprenticeship.

In the most impoverished sectors of the population the presence of images can be perceived where aspects of the cultivated culture are mixed with examples of an experienced culture which includes moral elements, personal experience elements and those of social behaviour associated with everyday objects. At all events, and in spite of the mention of aspects not purely academic, the majority of the comments transmit a detraction of the cultural practices of the groups to which they belong.

The academic representation of culture inhibits becoming aware of one's own practices as cultural, as well as inhibiting their appraisal as creative expressions of the social group itself. To the extent that it conceives of creation as an activity reserved for natural talents or selected minorities, it inhibits people from becoming aware of their own abilities for handling and transforming the reality around them. It becomes a factor inhibiting recognition and satisfaction of needs or participation in creation, re-creation, reflection about oneself, about others and the surrounding world.

To summarize, analysis of the needs and representations associated with the practices of the popular sectors shows:

(a) The gap existing between objective needs of participation, reflection, creation and re-creation and the population's subjective needs. The predominance of means of evading reality and the limited search for reflection, information and learning demonstrates a minimum recognition of these objective needs.
(b) The presence, in a small proportion of the population, of a certain recognition of their non-material objective needs.
(c) The presence of social representations which act as inhibitory factors, both for recognizing non-material objective needs involved in human development and for undertaking collective cultural practices favouring this development.
(d) The presence, in a minority, of images or visions related to the collective capacity for participation and transformer action, which can act as factors for recognition of needs of participation, reflection, creation and re-creation and of the emergence of productive-creative styles of action.

From a theoretical angle, this chapter thus assumes a dynamic interrelation between cultural practices, collective needs and social representations shared by a social group. The predominance of a consumerist or productive style in the cultural practices of popular sectors is associated with the degree of recognition of awareness of the need to be a protagonist in all areas of daily routine, and is rooted in a system of social representations which inhibit or facilitate this recognition.

We conclude this study with a quotation which refers to the idea of 'experienced' culture outlined above:

> It is necessary to give the word culture a real, experienced meaning. Culture is not only an equal right to consume certain values. Culture is not a mentality but rather the comprehension of the social context by individuals and groups. It means becoming aware of the values which exist in the social context and it is knowledge and awareness of self. It is the possibility of communicating with others, i.e. possession of the necessary languages for this communication. It is ability for self-expression by means of creative, spontaneous activities, whatever the level reached in the participation may be. It is social life experienced in the cultural sense of the word.

From this perspective of culture, popular and permanent education and action are not the result but the condition of social change.

Notes

1 For an amplification of the data presented in this chapter see M.T. Sirvent

and S. Brusilovsky, *Social-Cultural Diagnosis of the Bernal-Don Bosco Population*, Asociacion Cultural, Meriano Moreno, Bernal, Argentina, 1979.
2 The denomination 'social periphery' is generally applied to those cores of urban population which are most poverty-stricken in their economic and social living conditions. In this chapter social periphery is understood to mean the population excluded not only from economic benefits but also, and fundamentally, from the possibility of playing a leading role in facts affecting their daily life. In other words, they are excluded from participation in the spheres of work, family, community, mass media, education and free time. In this sense, all the popular sectors of Latin America are socially peripheral.
3 Data of the General Planning Coordination, Municipality of São Paulo.
4 It is assumed that the cultural area rests on a social, political and economic structure which acts as its roots and explains it. However, the initial assumption is that cultural aspects can operate as intervening variables in a transformation process facilitating or inhibiting the possibilities of this process.
5 Although we share an anthropological conception of culture understood as the form of life of a people, such as coherent articulation of beliefs, values norms and daily behaviour, in this chapter the whole universe of components of the culture of popular groups has not been taken; only those most relevant dimensions for the central concerns indicated in relation to a desirable model of human development and quality of life of the popular sectors are examined.
6 The word 'object' is used in a wide sense so that it covers material and non-material objects such as norms, values, or messages.
7 For further quantitative and qualitative data see M.T. Sirvent and S. Brusilovsky, op. cit.
8 The author of this chapter is currently co-ordinating several experiences of non-formal education and participative research in the urban area of Vitoria, capital of the State of Espirito Santo, Brazil. In this area it is common to observe groups of old women together with children looking for objects and food in the garbage deposits of the urban periphery.
9 The experiences of the author in the urban peripheries of the Brazilian cities show that the proportion of women who work outside home in a remunerated occupation depends on the degree of poverty. The lower the degree of poverty, the smaller the proportion of women with a remunerated occupation.
10 This occupation of bar, pool and football has been called 'the club of the periphery'.
11 The participation of the peace through union activity in the work has not been considered in this chapter, although it is a fundamental aspect of collective creation and transformation of popular working sectors of Latin America.
12 For more qualitative data on community social participation, see M.T. Sirvent and S. Brusilovsky, op. cit.; Secretaría Municipal de São Paulo, IICA, 1981.
13 SAB – Sociedades de Amigos del Barrio – are neighbourhood associations which were created in São Paulo, with the main purpose of facing

problems of material infrastructure such as water supply, electricity, pipes, gas and so on.
14 The preservation of these structures of solidarity and rural communication was found again by the author in later research in the urban periphery of other Brazilian cities.
15 We took this concept fundamentally from the study made by Rene Kaes about the representation of the culture of French workers. See R. Kaes, *Les Images de la Culture chez les Ouvriers Francais*, Paris: Cujas, 1968. Chapter 1 of the book is an excellent historical-theoretical presentation of the social representation concept. For further details, see also R. Kaes, 'Les Ouvriers Francais et la Culture', *Enquête*, 1058–61, Faculté de Droit et des Sciences Politiques et Economiques de l'Université de Strasbourg, Institut de Travail, Paris, 1962; P.H. Chombart de Lauwe, 'Hypotheses sur la Génesé et le Role des Aspirations et des Besoins dans les Societés du XXème Siecle', *Revue Philosophique*, 93e, 1968.

11
The African Personality

Bennie A. Khoapa

Any attempt to discuss 'personality' whether European, African, or Asian, is attended by a great deal of controversy depending on whether one seeks to understand it as a political, anthropological, sociological or cultural concept. Some attempts have been made to discuss it on the level of 'a way of governing' – which has been brought in by Western systems of economy and politics. Invariably, this approach has tried to show, however subtly, that the only history there is in Africa is that of the Westerners and that, therefore, the African Personality is at best the product of the 'civilizing effect' of Western culture on an essentially cultureless people on a 'dark continent'. Any evidence of an assertive position to counteract this arrogantly stupid way of evaluating Africa has been interpreted by these discoverers or writers, mostly Western, as 'pathology', 'predicament', or essentially proof of the African's basic inability to learn and comprehend what 'is good for him' – that good apparently being Western culture or civilization. A look at a great number of books written about Africa by Westerners will quickly make this point, however clumsily.

Another level – which is not accompanied by much writing – attempts to look at the African as he is, what is happening to him, what makes him 'tick'. This approach starts off from an acceptance of the fact that Africa has a people with a unique view of themselves and the world, in other words a people with a culture, inhabiting a continent with lots of sunshine and on the whole very little darkness.

This chapter takes the position that what defines the African Personality is his 'world view', i.e. his own conception of existential reality, his own view of his collective being or existence. It is this world view which in our view constitutes the primary lens through which he reduces the ineffable world of sense data to described fact. It suggests that one seeks to understand all else in the world to the extent that the rest of the world intersects and impinges upon his anticipation and control-needs, beginning with one's being and existence and then moving towards a universal understanding conditioned by that beginning.

It starts off with a look at the spring from which much of African thought comes. It attempts to show where the African Personality derives its 'soul'. That spring is African culture and philosophy. It makes allowance for the fact of impingement of other cultures and life styles upon the practical manifestations of this philosophical situation. It poses some problems resultant upon this impingement by other cultures on the so-called African way of life. It suggests that there have been many casualties along the way to a new selfhood, but that a strong resistance to a complete capitulation to those forces aimed at the destruction of this way of life is in evidence. This resistance draws its strength from the African philosophical roots or sociology of knowledge which I believe is the correct route to take on the way to discovering the African Personality.

This chapter will not pretend to provide the answers to the many problems facing the African in his attempts to be and remain himself. It will, however, allude to the general direction that is necessary to maintain the African Personality which, I believe, is a viable concept and is not, as some people think or wish, in decay. Endangered it may be, but dying it is not, because it is grown and rooted in rich philosophical soil.

We will hopefully develop a clearer understanding of the African Personality if we start from the beginning and deal with some description of key properties of African philosophy which have relevance to the subject: African Culture and the Concept of Man.

Culture

The generally accepted meaning of culture is that it is the sum total of ways of living built up by a group of people or human beings and transmitted from one generation to another. It includes, according to Sekou Toure

> ... all the material and immaterial works of art and science, plus knowledge, manners, education, mode of thought, behaviour and attitudes accumulated by the people both through and by virtue of their struggle for freedom from the hold and dominion of nature. We also include the result of their efforts to destroy the deviationist politics, social systems of domination and exploitation through the productive process of social life.[1]

It is wise to caution right at the beginning that human cultures resemble one another up to a point. Different cultures value their common elements differently only insofar as one puts the accent here and the other there. It is the ordering of the elements to one another via philosophical theories that determine the differences between cultures. Only the sum total of the particular ways of doing things, any one of which may also occur in other cultures, produces what can be called the 'uniqueness' which differentiates one people from another. The aggregate ordered by a particular philosophical conception constitutes the unique character of a culture.

When people choose a way of life, they also choose assumptions about 'man' that are consistent with that way of life. To the extent that there are differences in these assumptions on the nature of man, there are differences in cultures. In other words, societies have different cultures (personalities) to the extent that they have different assumptions or theories of human nature.

Concepts of man often differ fundamentally from society to society with resulting implications for political institutions and social control. What differentiates African culture from other cultures is the way in which the African views man and his position within society.

If, for example, we contrast the African view of man with that represented by one culture such as the Western liberal democratic culture coupled with one other type of Western culture represented by the Marxist view, we will hopefully help the reader to appreciate how the African perception of the role of man and state varies in some aspects so completely from these two. The two are chosen because of their relative influence over traditional African culture from the time of their first known contact with Africa to date.

It is because of the influence of these two cultures that the question about the validity or viability of the concept of 'African Personality' has arisen, with some people claiming that as Africa becomes more and more industrialized, it will also become Westernized, that is, deculturalized to an extent that to speak of an African personality will be 'romantic nonsense' at worst and misleading at best.

The Western liberal concept of man

An examination of all advocates of liberal democratic individualism reveals a consensus about man's dominant traits existing independently of any social context. The dominant notion in this conception being that individuals are '. . . abstractly given, with given interests, wants, purposes and needs . . . While society and the state are pictured as sets of actual or possible social arrangements which respond more or less adequately to those individual requirements.'[2] What is suggested by this notion is that individuals possess a realm of consciousness, beliefs or thought that does not affect others and that should remain immune from interference by external agents. In other words, the human personality has two realms – a social and a private realm – and society should have no control over the latter. One component of the 'private realm' is the 'inward domain of consciousness' which demands liberty of conscience which leads to the contention that matters of belief usually do not affect other people; therefore, beliefs should remain the sole concern of the believer and should be free from manipulation; an ideal which suggests that natural man (apart from society) is born with 'rights' which impose limitations on any societal restrictions to their pursuit.

People, according to this philosophical position popularized by European philosophers of the eighteenth century, have certain needs, interests and goals because they are the kind of creatures they are. Being rational, people discover that other people have the same needs and interests, and they learn God's rules whereby all can satisfy these interests. Learning God's rules, they are led to an understanding of rights. Each person has the right to require others to observe the rules in interacting with them. The 'rights' exist in the *state of nature* before human beings form societies.

According to John Locke, people enter this world with an interest in 'life, liberty and property'; and according to Thomas Jefferson, with 'life, liberty and happiness'. Life in this context refers to *self-preservation*. Liberty refers to the absence of restrictions on belief or expression of them – a right which is paramount in historical European democratic thought.

Big play is made by J.S. Mill of the concept of 'inner forces' in man which include the '. . . innate capacities and talents unique to each individual that can develop when not interfered with by outsiders.'[3] The ideal state of man then is associated with this psychological claim about capacities that are not socially induced. The development of these 'inner forces' is intrinsically desirable and also instrumentally useful as a means of achieving social progress. Therefore, according to this view, society should not interfere with their realization. They belong to man's individuality. Privacy as a value is closely related to the individual being left alone by other people to the extent that he can lead a life in a style and with satisfactions that give realization to the inner forces. Associated with this 'private self' is a third trait and value which assumes that people are capable of self-direction. Though not immune to social influences, they are nevertheless capable of reviewing their beliefs about values and norms that they and others have formed under that influence, critically evaluating them, and as independent individuals selecting or rejecting them and making choices for action on the basis of that review.

As Munro points out,

> This claim about the sanctity of the private realm is . . . construed from the concept of the dignity of man that is based on each person possessing a soul in the likeness of God. State policies that some

people regard as violations of autonomy through the manipulation of beliefs and choices are also viewed as violations of human dignity. This is because they seem to treat people as a means to state goals rather than as ends in themselves.[4]

The other Western view

Marxist theory claims that man's essential nature is social and that 'natural man' apart from social relations is nonsense. Marx makes an implicit distinction between 'man's social and biological natures', which assumes that when one talks about important aspects of man, one is talking about his social nature. He repudiates the existence of any innate drives to sympathetic conduct in man. In his Sixth Thesis on Feurbach, he talks of the human essence being '. . . no abstraction inherent in each single individual. In its reality it is the ensemble of social relations.' These social relations he views as '. . . legally defined links in co-operative production (of goods other than those needed for biological survival) that exist between the dominant groups in any historical period: slave-master, serf-lord, employee-employer.' Thus he repudiates the notion that man's social nature is a static substance; rather suggesting that it is something that changes as the relations change.[5]

An important point which Marx makes is that man is the only creature for whom social relationships take on a class form, and that man uniquely fulfils his social essence or becomes a social being through labour – a position that echoes a point of Hegel. Other animals may get goods needed to survive from what is around them, but man, through his productive labour creates these goals. His singular social and economic activity satisfies his needs and realizes his potentialities. In sum, Marx and Engels' account of human nature is that it is determined both by what men produce and how they produce it. New developments in technology occur continually, which means that how men produce is always evolving. Thus, gradual changes in human nature are inevitable.

The African concept of man

The African view of the world is one of harmony. This harmony which demands mutual compatibility among all the disciplines considered as a system constitutes the main basis of African thinking. Manifest in the most rudimentary elements of nature is God. Philosophy, theology, politics, social theory, land law, medicine, birth, burial – all find themselves concentrated in a system so tight that to exclude one item from the whole is to paralyse the structure as a whole.

The African's conception of man sees biological life and spiritual life *meeting* in the human being and neither the one nor the other being present alone. The essence of human life is the unity of both principles. Man shares biological life (natural life) with the animal, but spiritual life divides him from the animal and gives him his 'Personality'.

So, it is fair to say that the African view of the universe and of man within that universe is profoundly religious. They see it as religious and treat it as such; and, while there are many different accounts of the creation of the universe among African people, we can generalize that they all put man in the centre of this universe. As it is often said '. . . Man . . . who lives on earth, is the centre of the universe. He is also like the priest of the universe, linking the universe with God, its Creator; Man awakens the universe, he speaks to it, he listens to it, he tries to create a harmony with the universe. It is man who turns parts of the universe into sacred objects, and who uses other things for sacrifices and offerings. These are constant reminders to people that they regard it as a religious universe.'[6]

A look at one cultural institution in African society will make the point – the institution of marriage. Man, according to African thinking, is constructed for reproduction. To leave no living heirs behind him is the worst evil that can befall a man and there is no curse more terrible than to wish an African man or woman to die childless.

For this reason, in African society, bearing children is an obligation, and that obligation is fulfilled through marriage. Failure to get married is like committing a crime against traditional beliefs and practices, a point which is often missed by Westerners when they import into Africa their concept of family planning.

Marriage is the uniting link in the rhythm of life. All generations are bound together in the act of marriage – past, present and future generations. 'The past generations are many but they are represented in one's parents; the present generation is represented in one's life, and future generations begin to come on the stage through childbearing.'[7] Since the supreme purpose of marriage, according to African people, is to bear children, to extend life, a marriage becomes fully so when one or more children have been born.

Marriage also provides for new social relationships to be established between the families and relatives involved. It extends in other words the web of kinship socially. Marriage is above all intimately linked with the religious beliefs about the continuation of life beyond death. Through marriage, the departed are in effect 'reborn' not in their total being but by having some of their physical features and characteristics or personality traits reborn in the children of the family.

Marriage gives a person – according to African thinking –

'completeness'. It is part of the definition of who a person is according to African views about man.

Children, it follows, are greatly valued in African life, for they are the seal of marriage; it does not matter whether one is talking about traditional or so-called 'modern Africa'. Children are believed to prolong the life of their parents and through them the name of the family is perpetuated. Therefore, the more a person has, the bigger is his glory.

Children add to the social stature of the family, and both girls and boys have their social usefulness in the eyes of their families. At home, there are duties which the children are expected to do as their share in the life of the family. They are taught obedience and respect towards their parents and other older people. They help in the work around the house and in the fields. As they grow older, they gradually acquire a different social status and their responsibilities increase.

When the parents become old and weak, it is the duty of the children, especially the heirs or sons, to look after the parents and the affairs of the family.

For African people, the family includes children, parents, grandparents and other relatives such as brothers, sisters, cousins, uncles, etc. All relatives have duties and responsibilities towards one another. Everyone knows how he is related to other people in the clan and in the neighbourhood. This idea of family extends to include the departed as well as those still to be born. John Mbiti makes this point strongly when he says:

> In African life, we cannot speak of marriage alone, it is always in terms of marriage and family life. One gets married within the context of family life and one gets married in order to enlarge that family life. One stands on the roots of family life; and one puts out branches of family life. This idea of the individual in relation to marriage and family life is deeply rooted in African thinking.[8]

Human conduct: an African view

It is, perhaps, in this area that there are important distinguishing differences between the African and the Western man and we will do well to spend some time examining it, as it has important implications for social, political, economic differences between these societies. Morals in all cultures are central to human conduct, as they deal with the question of what is right and good and what is evil and wrong in human conduct.

It is the moral sense of a people that produces customs, rules, laws, traditions and taboos which can be observed in each society. African societies believe that their morals were given to them by God. This view provides an unchallenged authority for morals. It is also thought that some of the departed keep watch over people to make sure that they observe the moral laws and are punished when they break them.

Human conduct has two dimensions in African life. There is personal conduct, which has to do specifically with the life of the individual, for example, whether he should help in the field; or whether he should buy clothes for himself and his family. The greater number of morals has to do with social conduct, i.e. the life of society at large, the conduct of the individual within the group or community or nation. African morals lay a great stress on social conduct, since a basic African view is that the individual exists only because others exist. Moral laws help people do their duties to society and enjoy certain rights from society. Morals are what have produced the virtues that society appreciates and endeavours to preserve, such as friendship, compassion, love, honesty, justice, courage, self-control, etc. On the other hand, morals also sharpen people's dislike and avoidance of vices like theft, cheating, selfishness, greed, etc.

Many morals have become rooted in the life of the peoples concerned, in Africa, no less than in other cultures.

African family morals

Each person in African traditional life lives in or as part of a family. Kinship is very important in all aspects of African life. The family is the most basic unit of life which represents, in miniature, the life of an entire people. In the family, individuals are closely bound to each other, both because of blood or marriage and because of living together.

In all African families, there is a hierarchy based on age and degree of kinship. The oldest members have a higher status than the youngest. Within that hierarchy there are duties, obligations, rights and privileges dictated by the moral sense of society. Parents have distinct duties towards their children such as the duty to bear the children, protect them, educate them, discipline them and bring them up to be well behaved. The children on the other hand have clear duties towards their parents such as obeying their parents, doing as they are told, assisting with family chores, respecting those who are older, being humble in the presence of their parents and other older people and of course much later when they are older to take care of their parents when they are old and sick.

If parents fail in their duties towards their children, the wider communities may punish them. If the child fails in his/her duties, he/she will be punished. At home, it is expected that children will learn to tell the truth, to help others, to be honest, generous, considerate, hardworking, etc.

There are also morals concerned with hospitality to relatives, friends and strangers. It is held to be a moral evil to deny hospitality, even to a stranger. When people travel, they may stop anywhere for the night and receive hospitality in that homestead.

Other family morals concern property, the care of the home, the fields, the animals.

Community morals

What strengthens the life of the community is held to be good and right. What weakens it, it follows, is held to be evil and wrong.

There are morals concerning the social, economic and political life of the people as a whole. These cover aspects like mutual help in time of need, maintaining social institutions like marriage and the family, defending the land in time of invasion or aggression. Many other things are held to be morally wrong such as robbery, murder, rape, lying, stealing, showing disrespect, interfering with public rights, breaking promises and so on. All these and many others are moral vices in the eyes of the community.

There are also many things that are considered to be morally right and good such as politeness, kindness, showing respect, being reliable.

The preceding material can and is often challenged as being inadequate to describe the modern African, who it is argued, has undergone tremendous change as a result of the impact of other religions and cultures in Africa. Also it is often pointed out that the diversity of African people defies any attempt by anybody to profile an 'African Personality'.

Impact of other religions and cultures

Yes, the preceding material attempted only to give a backdrop to the African personality because it is important to understand where the so-called 'modern African' comes from. It would not be possible to understand him if we did not understand this background which springs from traditional African culture. We now turn to the African after the arrival of other religions and culture in Africa.

First, we will deal with the arrival and the impact of Islam on Africa and consider later the arrival and impact of Christianity on the scene. From there we shall attempt to figure out how the African personality has been affected by these tremendously important events in the history of the continent.

Islam and Africa

If the peoples of North Africa are included, more than 100 million Africans embrace Islam today; some claim that the numbers are increasing rapidly. They are more numerous than the followers of any other organized religion in Africa. Why has it proved so attractive, comparatively, even though '. . . they came to conquer not by the doctrine of love which Christianity at its best teaches, or by persuasion, but by the sword'?[9] Some people suggest that Islam is 'relatively tolerant of non-believers . . . it puts greater stress on the brotherhood of man than on the need for all men to worship alike.' Some are not so sure. Generally speaking, it is suggested that the Muslims aroused less resentment among Africans than did the Christian missionaries, many of whom taught that the Christian way of life was the only acceptable one for all human beings. Nor were Muslims, it is suggested, as concerned about race as many Christians seemed to be. The long contact of Arab and Berber merchants with African peoples was also a significant factor in reducing feelings of 'differentness' between the two races, some suggest.

Christianity

There is also evidence that African peoples are responding to Christianity in spite of its having been associated with colonial rulers. Estimates indicate that there were about 165 million in Africa in 1974, and according to Mbiti their numbers were increasing at the rate of 5% every year. If that rate is maintained, it is estimated by some that by the year AD 2000 there will be roughly 400 million Christians in Africa.

Christianity has had a major impact on the lives of African peoples. It has built schools and hospitals at which the majority of African leaders of today were educated. 'It is also by the ideals of justice, human dignity, love and brotherhood that African leaders were inspired to fight against colonialism and foreign domination.'[10]

The biggest obstacle to the spread (further spread) of Christianity is that in many people's minds it is still associated with the arrival of colonialism in Africa and all that is associated with it – a matter we shall visit later.

The so-called modern Africa

Since the advent of European culture, through the colonialization of the African continent, there has been an attempt to bring about disassociation in African life between a way of life and a way of governing.

We must now return to the main purpose of our discussion, which is to reflect on the African personality in the midst of the great confusion which has brought about debates around questions like 'Is there such a thing as an African Personality?' In considering the question of the African Personality, we have of necessity to consider the impact on African culture of Islam and Christianity. The extent to which Africans have embraced both Christianity and Islam makes some wonder whether there is enough of an 'African Personality' left to justify our speaking of it in terms of an autonomous 'self'. Great debates on this subject have been heard from Nkrumah to Nyerere – perhaps more so in the context of 'Pan African' debates which received general attention in the years following the independence of Ghana in 1957, reaching a peak in the 1960s when most of Africa gained political independence.

No one would deny that the literal bombardment of African people with the cultural arrogance, particularly of the Western colonizers – coming often in other guises such as 'scholars', anthropologists, missionaries, historians, etc – has indeed left large footprints of ruins and scars all over the continent. These are scars which do not heal very quickly mainly because they are deeply imprinted in the brains of the victims.

Yet, we see evidence of a slow recovery from this 'assault'. This recovery is manifested by evidence of the continuing survival of traditional traits in African society. The sense of community, the rituals surrounding birth, marriage, death, the theatre that surrounds African life in general, are all traits that have defeated any possible bourgeois desire to be left alone. More importantly, there is evidence that these remaining traits are being forcefully translated into the formation of political and educational institutions as well as economic programmes that show that the African Personality is alive and on its way to recovery.

Westerners still get surprised when they observe Africans relating to each other in the course of normal intercourse – the laughter, the spontaneity, the apparent no end to talk, whether in formal or informal meetings. To Africans, the mere being together seems sufficient reason to engage in exchange. They do not wait until introduced to each other as one often sees in Western culture where one has to be introduced before he can engage another person in conversation. The deafening silence which one observes when travelling in the Western world is not only confusing to Africans but is also very strange.

To the African, every gathering of people is simply an extension of the family and thus demands no inhibitions of behaviour that one would have displayed in a smaller group.

Similarly, with home visiting, no reason is necessary for an African to 'drop by' and engage in conversation. Contrast this with the Westerner's way of life where a visitor to someone's door is met with something like 'Can I help you?' – suggesting that unless you have a 'good reason' for stopping by, 'you had better get going and leave me in peace.' In Western society, it seems any social visit must be by 'appointment', otherwise it constitutes discourtesy and a violation of the 'personal rights' of the other person. It must have been anger and surprise at this type of behaviour on the part of the Westerner which prompted Mphahlele to write recently:

> . . . I have not yet seen an African explore territory or climb mountains for mere conquest; I have not yet seen him sit on a lonely rock or river-bank or lake fishing; I have not yet seen him develop game parks except what he inherited from colonialism. It is a silly Western idea to conserve wild life for the entertainment of foreign tourists. Yes, it brings in revenue, but, left to ourselves, we would rather use the land to provide food for us . . . I have no sympathy with white people who weep over the disappearance of a species of lion or elephant or leopard; because this Western attachment to animals, wild and domestic, that goes beyond basic utility disgusts me, it does most other Africans . . . I have not yet seen an African go out to lonely places for a vacation, just for the scenery. We go to other people.[11]

Only a person who enjoys talking to other people would 'go to other people'. Talking, or discussing, is one thing which the African can do – whether at work, at school, in government or in church. Nyerere commented once, 'It is a rather clumsy way of conducting affairs, especially in a world as impatient for results as this of the twentieth century, but discussion is one essential factor of any democracy; and the African is expert at it.[12] Would we then trade this 'clumsy way' of discussing for the so-called 'civilized, efficient way represented by Western liberal democracy'? The answer would be: no need; democracy is not anything new to Africa, we need no new type. In fact, Nyerere refers to this idea when he says

> . . . these three . . . I consider to be essential to democratic government: discussion, equality and freedom – the last being implied by the other two. . . . The traditional African society . . . was a society of equals and it conducted its business through discussion – 'they talk till they agree' so the saying goes – That gives you the very essence of traditional African democracy.[13]

There is another reason why the kind of 'democracy' which Westerners believe we should rather be following is unacceptable to us. It is because of the contradictions in it which to us at any rate abound. When we read a massive array of literature on Western democracy, we are amazed by these contradictions. The very founders of Western democracy – the Greeks – could talk of 'democracy' even while more than half of their population had no say in the conduct of the affairs of state. The founders of the American (US) nation could talk about the 'unalienable rights' of man even though they believed in exceptions. Their most illustrious leader, Abraham Lincoln, could bequeath to us a perfect definition of democracy although he spoke in a slave-owning society. The British could brag about 'democracy' and still build a great empire for the glory of Britons. These nations believed in government by discussion, by equals, but lived in a world which excluded masses of human beings from their idea of 'equality' and felt no scruples about it.

Developments in African countries are a clear indication that Africans have grave doubts about the suitability for Africa of the Anglo-Saxon form of democracy. In his own traditional society, the African has always been a free individual, very much a member of his community, but seeing no conflict between his own interests and those of his community. This is because, as we have mentioned earlier, the structure of his society was a direct extension of the family. First, there was the small family unit; this merged into a larger 'blood family' which in turn merged into a tribe. The affairs of the community were conducted by free and equal discussion, but nevertheless, the African's mental conception of government was personal – not institutional.

Yet, Africa still has problems which prompt some people to say that an African personality is a myth and that sooner or later Africans will come to realize that progress can come to them and their countries only when they forget about the 'sentimental talk' and acquire Western ways of doing things. 'Either Africa adopts the Soviet way of doing things or it adopts the Western way of doing things or it sinks' seems to be the constant suggestion from priest to politicians of the West and East. According to these people Africa must choose between the horns of a dilemma.

Although part of the problem has been that this kind of propaganda has found some support in some African circles, it has also worried many other thinking people and has consequently come close to developing a predictable kind of 'disease' which Mazrui calls 'cultural schizophrenia' found in the growing class of so-called educated élites, which must constitute the most difficult of all problems as it strikes at the root of the question of relevance.[14]

It is worth repeating that the colonization of Africa was not only a political experience, it was much more a cultural experience. The values of the African that we talked about earlier were seriously disturbed by this experience.

To understand how this acculturation problem came about, it is important to examine the role of 'education' as brought in by the colonizer. The inherited colonial education had as its principal objective the de-Africanization of African nationals. It was discriminatory, mediocre and based on verbalism. It could not and (as it is increasingly being realized by African leaders) cannot contribute to national construction or reconstruction because it was never intended to achieve that purpose. Divorced as it was from the reality of the people, it was for that reason a schooling for a minority and consequently against the majority. It selected only those few who had access to it, excluded most of them after a few years and due to continued selective filtering, the number rejected constantly increased. A cursory look at the trends in African education in all countries that were colonized will show this. Just one example would be that of the Belgian Congo – now Zaire. When that country attained independence in 1960, there were less that 20 African people with a university education after a period of over 100 years of Belgian colonization. Other examples, not quite as bad as this one but still frightening are found all over Africa. A sense of inferiority and of inadequacy was fostered by this 'failure'. This kind of education in other words spread a 'culture of failure'; to miseducate, if you like.

This was in line with the colonizers' conception of the people of the continent as inferior people, and so the system produced in African children the profile that colonial ideology itself had created for them, namely, that of inferior beings, lacking in all ability. In this way, the continued presence and domination of the colonizer over the colonized was justified. Lacking this ability, the system suggested to Africans that the only way to acquire the 'ability' was to become 'white' or 'Western'. This system of education was not concerned with anything closely related to the nationals; worse than the lack of concern, however, was the actual negation of every authentic representation of national peoples – their history, their culture, their language. The history of those colonized was thought to have begun with the colonizing presence of the colonizers. The culture of the colonized was a reflection of their 'barbaric' way of seeing the world. Culture belonged only to the colonizers.

The paradoxical thing, though, is that while the education given by the colonizers was intended to consolidate colonial control, it also contributed to the arrival of African nationalism, expressed by the clear African rejection of this obvious attempt to poison the minds of the people of Africa. These Africans rightly decided to assume their own history, inserting themselves into a process of 'decolonization of their mentality'. This process resulted in the political decolonization which gathered momentum in the 1960s, setting off in the process a general reawakening of black people in other parts of the world, notably the USA.

Political decolonization became the first manifestation of what came to be referred to as the African Personality. In more recent times, it has taken the form of an assertive position of radically transforming, amongst other things, the educational systems inherited from the colonizer. It has assumed political decisions coherent with the plan for the society to be created or recreated and is based on certain material considerations that offer incentives for change. It demands additionally, increased production; at the same time, it requires a reorientation of production through a new concept of distribution. It says that a high degree of political clarity must undergird any discussion of what to produce, how to produce it, for what and for whom, in the spirit of African culture.

What gives birth to and nurtures a personality is the socialization process or education. Therefore, the act of educating, its possibilities, its legitimacy, its objectives and ends, its agents, its methods and its content address the primary factors involved in the development and presentation of a culture. Consequently, when we talk about what should be known, we also must involve the question of why it needs to be known, how and for whose benefit and in whose interests.

Julius Nyerere in his book *Education for Self-Reliance* makes this point when he talks about education being 'preparation for life' which consists of a critical understanding of the life actually lived; for only in this way is it possible to create new ways of living. His thought, which is both pedagogical and political, is nourished by what is real, concrete and based on experience, the transformation of which is the central educational activity.

What this says is that defining what to know demands political clarity of everyone involved because this clarity, as Paulo Freire points out 'is essential to the answering of all other questions that follow, such as, for what reason and for whose interests do the policies themselves serve?'[15]

The knowledge of how to define *what* needs to be known cannot be separated from the *why* of knowing. For this reason, I agree with Freire that '... there are no neutral specialists, there are no neutral methodologists who can teach how to teach history, geography, language or mathematics neutrally.'[16] This is illustrated by this example of a British history professor who is said to have replied thus to a student who demanded more content on African history in his class:

> ... Perhaps, in the future there will be African history to teach; but at present there is none; there is only the history of the Europeans in Africa. The rest is darkness ... and darkness is not a subject of history.[17]

If we accept the statement made by this 'historian', then we do not need to discuss the matter any further, for we should simply conclude that there is no such thing as African history, African culture and therefore by implication no such thing as an African personality.

But, of course, I do not accept this arrogant statement and neither do most Africans. Despite the fact that in the process of development, African countries have taken the risk of accepting certain Western models in their attempt to deal with the problems of reconstruction, among which was an educational system produced in a class society, there are African leaders now who are increasingly discovering that it is not possible to develop national leaders to undertake the enormous task of national reconstruction with only middle-school and university degrees obtained in the Western model, and consequently attempts are now being made to find the correct route to combating the anti-popular leadership style that surrenders to the interests of foreign imperial culture.

Admitting that the colonizer offered some opportunity for persons to gain university qualification, this action was nevertheless taken by the colonial power in its own interest. What was offered was selective and narrow, like the type of education that evolved in the urban schools of the colonial school era. Reaching only a small segment of the population, this university training reinforced the ranks of urban intellectuals in the service of the colonizer – creating this serious problem of split personality (schizophrenia) or élitism.

When African leaders like Nyerere and the late Amilcar Cabral and others call for the need for such 'intellectuals' to 'commit suicide' and be 'born again' as leaders entirely identified with the most profound aspirations of their people, they imply the re-Africanization of these intellectuals.

It so happens, however, that committing suicide is not pleasant and it has not been easy for these 'intellectuals' to accept this 'necessary death' as evidenced by the migration of élites from those countries which make demands on the educated élites to bring their conduct more in line with the aspirations of the people. This is because the intellectual training of the middle-class reinforces the class position of individuals and tends to make them absolutize the validity of their own activity, which they consider superior to that of those without the same opportunity.

It becomes a tragedy in my view when an African country fails to revolutionize the content and structure of education, when it fails to institutionalize culture to help re-educate the élite that already exists, for surely the challenge for such a society is not to continue creating élitist intellectuals so that they can commit the kind of suicide that is called for, but rather to prevent their formation in the first place. The re-orientation of the educational system is the only way by which Africa can totally overcome the colonial inheritance. It demands different objectives, different content, different practice and a different conception of education. To discuss education is to think of the overall plan for society itself including an economic system.

African socialism

If, therefore, African society seeks to remake itself, move forward towards a satisfactory care of its inhabitants, it needs on the one hand to organize its methods of production with this objective in mind and on the other to structure its education in close relation to production, both from the point of view of the understanding of the productive process and also the technical training of its people.

Tom Mboya is not the only one who talked about a strong belief that in 'economic relations we can similarly be guided by the traditional presence of socialist ideas and attitudes in the African mental make-up.'[18] It was left to Nyerere of Tanzania, however, to boldly espouse this concept further and have the courage to make it the cornerstone of his political policy.

When reference is made to socialism, there is always an immediate response from those Westerners or Western-educated Africans who think of socialism of the Western type, e.g. that of Britain or Sweden. Others will also think of the Marxist brand of socialism. The kind of socialism we talk about here is African socialism, which refers to those codes 'of conduct in the African societies which have, over the ages, conferred dignity on our people and afforded them security regardless of their station in life. I refer to universal charity which characterized our societies and I refer to the African's thought processes and cosmological ideas which regard man, not as a social means but as an end and entity in the society.'[19]

Both Tom Mboya and Nyerere make the important point about the socialism we talk about and this is that the basic tenets of socialism are universal; socialism did not start with Marx: he did contribute a great deal to social thought; he did not, however, invent socialism. In Western socialist tradition, for example, this 'mental attitude' began with Aristotle's dictum that man is a social animal which has no potency and no life outside the society. From this has arisen a host of economic, social and political thought amongst which was capitalism.

In a capitalist perspective, the various factors in production – means of production on the one hand, workers on the other – combine in the service of capital. Part of the accumulated profits which are not paid to the worker who sells his labour to the capitalist, is used for the capitalist's well-being. Another part is used to buy more labour and more means of production which, together, produce more goods to be sold. The capitalist is interested in the production of goods – not, however, in terms of their usefulness, but rather in their value as a means of exchange, that is, goods that can be sold. What is more, he seeks to produce goods whose value covers and surpasses the sum of the values of his investment in production – the means of production and the labour.

What workers receive as salary for their effort expended in the act of production corresponds only minimally to their effort. What is available for their living is also minimal and, therefore, the wage-earner class reproduces itself.

Deprived of the product of his labour, the worker has no say in the determination of what will be produced. To the degree that a significant quantity of what is produced does not correspond with the real needs of individuals, it is necessary to invent needs. Thus, a society becomes totally ambiguous when, in attempting to follow socialism, it allows itself to become fascinated by the myth of consumerism. If it moves in this direction, even though it does not have a capitalist class, its objective will be to produce goods to be sold. The socialism we are talking about is something quite different from a 'capitalist society without capitalists'.

A capitalist society is a society of consumption. The role of advertising in such a society, with its alienation of the conscience is fundamental to this kind of society. We do not need advertising to convince our people to buy mealie-meal, rice or yams; but we do need advertising to buy this or that type of perfume or that kind of rice if the only differences is in the packaging. If production is governed by the well-being of the total society rather than by capitalist, private or state, then the accumulation of capital – indispensable to development – has a totally different significance and goal. The part of the accumulated capital that is not paid to the worker is not taken from him, but is his quota toward the development of the collectivity. And what is to be produced with this quota is not goods defined as necessarily saleable, but goods that are occasionally necessary. For this it is essential that a society re-constructs itself to become a society of workers whose leadership renounces both the tendency to leave everything to chance and the hardening of bureaucracy.

The more a society is capitalist and therefore class conscious, as well as without political consciousness of individuals with regard to the recreation of the society as it moves toward becoming a society of equal people, the more wholeheartedly the people give themselves to productive efforts. Their political consciousness is a factor in their attitude toward production.

In a capitalist society, the education of workers has as one of its goals the continuation of a class of wage earners, obliged to sell their labour to the capitalist class. The education required to continue reproducing this class is one that will continuously increase efficiency of the workers in their participation in the work process.

The foundation, then, and the objective of African socialism is the extended family. The true African socialist does not look on one class of men as his brethren and another as his natural enemy. He does not form an alliance with the brethren for the extermination of the 'non-brethren'. He rather regards all men as his brethren – as members of his ever-extending family. The concept of 'Ujaama' which speaks of 'familyhood' comes closest to describing the socialism that we speak of.

It is opposed to capitalism which seeks to build a happy society on the basis of the exploitation of man by man and it is equally opposed to doctrinaire socialism which seeks to build its happy society on a philosophy of inevitable conflict between man and man. In a society seeking to reconstruct itself along these socialist lines, education should be pre-eminently revealing and critical. Generally the need is to overhaul the whole structure so that what subject is taught or not taught, the order of importance and priority assigned for it if it is taught, its content, will be determined by African national philosophies of education. A government should be able to say: this is our education because this is our culture; these are our national standards; this is how we want to interact with the rest of Africa at one level and the rest of the world at another; from here we can understand what it is we want from societies outside ours.

The African intellectual

Is he consistent with the African personality? Perhaps that is debatable, but what we do know is that a new generation of Africans is emerging that realizes that the headlong embracing of other people's experiences without first having mastered their own intellectual responsibility is dangerous. This realization is also part of the search for a new way, part of the recognition of the utter bankruptcy of 'bourgeois' thought and part of the desire to replace it with something more relevant.

There are Africans today who do intellectual work. This is neither a cause for shame nor celebration. There is a role to play for such intellectuals in the struggle to maintain the continent's self-respect. The task is to identify that role and play it.

The new African intellectual realizes that his opportunity for self-realization as a group can only come through an alliance with his people, that even though one may choose to be an individual scholar, no matter how formidable the mind and overwhelming the productivity, no single individual, regardless of how gifted, can change a social system by himself. Those African intellectuals who have lost their way have done so because they have tended to substitute their reality for the world's and because they have estranged themselves from their people – their only valid frame of reference.

The real problem has finally (even though slowly) been discovered – that much of our past intellectual difficulty has resided in trying to create cultural and political theories divorced from the struggle to re-Africanize Africa fundamentally and irrevocably; that the problem of creating the necessary theory and then integrating it with the struggle for state power can only take place in conjunction with the masses of the people (the African people as they are) and as part of the process of correctly identifying their problems and the solutions to those problems.

We must agree with Franz Fanon when he says '. . . no one can truly wish the spread of African culture if he does not give practical support to the creation of the conditions necessary for the existence of that culture.' Fortunately, a breed of African is emerging that will demand institutions of culture that represent a real African consciousness – the African Personality.

Notes

1. Sekou Toure cited by Leroi Jones in *Black Value System*, UCLA.
2. Steven Luke, *Individualism*, Oxford: Basil Blackwell, 1973.
3. John Stuart Mill, *On Liberty*, New York: Library of Liberal Arts, 1956.
4. Donald Munro *The Concept of Man in Contemporary China*, Ann Arbor, MI: University of Michigan Press, 1977.
5. Marx & Engels in Lewis S. Feuer (ed.) *Basic Writings on Politics and Philosophy*, New York: Anchor Books, 1959.
6. J.S. Mbiti, *Introduction to African Religion*, New York: Praeger Press, 1975.
7. Ibid.
8. Ibid.
9. Ibid.
10. Ibid.
11. Z.K. Mphahlele, *African Images*, New York: Praeger Press, 1974.
12. J. Nyerere, 'The African Democracy' in James Duffy and Robert Manners (eds.) *Africa Speaks*, 1961.
13. Ibid.
14. Ali Mazrui, *Political Values and the Educated Class in Africa*, Los Angeles: University of California Press, 1978.
15. Paulo Freire, *Pedagogy in Progress: Letters to Guinea-Bissau*, New York: The Seabury Press, 1978.
16. Ibid.
17. Hugh Trevor Roper, *The Rise of Christian Europe*, London: Thames & Hudson, 1966.
18. Tom Mboya, 'African Socialism', quoted from *Transition* Kampala, 1963.
19. Ibid.

12
Buddhism and Development

Sulak Sivaraksa

The Venerable Buddhadāsa Bhikkhu, a famous Siamese Buddhist monk, once remarked that the word 'development' in its Pali or Sanskrit equivalent means 'disorderliness' or 'confusion', and in Buddhism 'development' refers either to progress or regress. In a similar vein, Ivan Illich once told me that the Latin word 'progressio', which is the root idea of 'development', can mean 'madness' also.

When we look at what is happening in the world today, where national development is being so much emphasized, especially in the so-called underdeveloped or the developing world, it is difficult to contradict what these two religious leaders were implying. The further development has proceeded, the more it has resulted in the rich becoming richer, while the poor remain poor or become even poorer – and the rich are still not happier. Nature and the environment are deteriorating day by day. Animal life and other natural resources are increasingly wasted for selfish purposes to such an extent that some of the world's leading professionals, who organized the Club of Rome, have prophesied that unless mankind makes some fundamental changes in its world view, continued development may result in the ruin or destruction of the world within this generation or the next. Is this overly pessimistic?

In fact development may be viewed from either of two aspects: quantity or quality. This does not mean that the two are mutually exclusive, for quantity and quality are quite inseparable. But the question is, which one is to be viewed as basic and which one as secondary.

Accepting quantity as the basic factor, one can measure the results of development in terms of physical things, things that can be measured or touched such as increase in income, factories, schools, hospitals, buildings, food, clothing, the labour force, etc., assuming that these are all good things, and that the more the quantity increases the more the quality automatically is increased. But the question is, is this a correct assumption? As I have suggested, it is now clear that it is not correct.

Accepting quality as the basic factor, one can measure the results by looking at human beings themselves. But one must view people in relation to their full potentiality. It is not correct to say that when people have enough to live on without starving and are free from disease, that they have sufficiency, even though they may lack freedom and liberty; or to say that it will be quite satisfactory if only *most* of society can be developed in a material or quantitative sense. Is it not true that those who have become excited about the successes of China or Russia have not considered people in their wholeness? I do not want to belittle those two countries, especially China, which has developed so far materially. We must admit that China has had no small success. But if you say that such societies have also been successful in the matter of the quality of human life, I would have to disagree. Most of those who praise such societies have no desire to go and join them, unless there is no other option. To put it another way, development through socialistic communism only becomes possible when there is no other way open, for people want to realize their full humanity. The four Buddhist 'requisites' (food, clothing, shelter and medicine) help only to keep one alive. But people want and need to go further and search out their own abilities as far as possible, and then to use these hidden abilities to the fullest extent, through social or institutional structures for the general welfare. To put it in religious terms, this has to do with the sacredness or the special quality which man has. In this view, development must at every step take into account the essence of humanity. It must deal with the question: 'What is man, and what should man be?' Development in this sense will not overlook the need for the four requisites, for they are necessary for life itself. But they are not sufficient. Emphasis should not be put on luxuries beyond the four requisites, while some fellow humans still do not have enough of the four requisites to sustain life. Emphasis should be placed on the *quality* of human life, moving beyond the supplying of the bare necessities of living, to bring human life its highest fulfilment.

The problem in qualitative development lies in the difficulty of identifying and measuring qualitative results. Planners on the whole, whether in the East or the West, do not want to take the time to study the complicated problems having to do with ultimate goals of humanity. They excuse themselves by saying it is a problem of metaphysics or of religion, as though ultimate goals are beyond the ability of common people to discuss or understand. Philosophers and theologians or religious leaders are not without fault in this, as they do not give sufficient attention to using language that common people can understand, or they show obvious disdain for non-religious people, while at the same time they evidence little interest in the development plans of experts in other fields. Or they simply occupy themselves with things that have little to do with the life and death matters of people. They spend their time on trivia and on the outward forms of religion, emphasizing this or that aspect of theology or of ceremony, and some even make their religion into a business.

Because of the failure of development to take quality into account due to the difficulty of assessing results, and because of the refusal to examine the essence of true humanity, development generally is simply a matter of quantity. The greater the development in society, the more complex society becomes, until development requires people with highly specialized knowledge who find it difficult to be interested in the wider problems of humanity. Even their language becomes so technical that only their own inner circle can understand it. (In this respect those who set themselves up as religious experts are no different from other experts.) What is worse is that it is possible for these experts to measure results in terms of their own success, in their own profession, with scarcely any consideration given to whether or not society has been bettered or not. For instance, success in religion is measured in terms of the number of churches or temples built, their beauty, the income of the religious organization, increase in membership, and the quantity of material published, without necessarily taking any note of possible change in the basic humanity of character of the members, their self-sacrifice, or their true neighbourly love, not to mention any improvement or deterioration in matters of the mind or purity of spirit. In the field of educational development, success is spoken of in terms of the increase in the number of students and schools, and the expansion of curriculum, with no method of determining to what extent, if any, the studies are helpful to the students. In the medical field, what seems to count is the number of doctors and hospitals, or the public health budget. This or that country is held up as an example of medical leadership because of heart transplants or kidney transplants, or the prevention of this or that disease, without taking note of the fact that in that country everyone has become a patient and no one has any ability to take care of himself. They have become totally dependent on doctors. A simple headache or a slight fever, and they must rush to a clinic or the hospital. Even an argument between a man and a wife requires the services of a psychiatrist. Moreover, it can be asked, to what extent is modern medicine responsible for the population explosion? In the field of transportation, success is measured in terms of speed. Some people have to spend most of their time travelling, while many more people do not even have a chance to leave their village. Yet there is no consideration given to the question whether it is beneficial or detrimental to leave one's village or to be able to go where one wants to go.

Even worse is the fact that economists measure success in terms of increased production, and so must turn to industrialization, to the profit motive. And it is the economists who are most influential in development planning. As a result development becomes supremely a matter of economics and politics. For economists see development in terms of increasing currency and things, thus fostering greed (*lobha*).

Politicians see development in terms of increased power thus fostering ill-will (*dosa*). Both then work together, hand in glove, and measure the results in terms of quantity, thus fostering ignorance (*moha*), and completing the Buddhist triad of evils. This is illustrated by the fact that almost every country aims to increase the gross national product, to increase the trade balance, to increase exports, to expand its industry, to expand building construction, etc. As for *people*, they are considered only as the labour force and as consumers. So people have value only as a means to make the numerical statistics of success look good on paper. No consideration is given to the fact that those people must endure tyranny, and are taken advantage of, while nature around them and their own style of living deteriorate. Take as an example the construction of a new industrial plant in a developing country for which modern machinery is brought in from abroad, with foreign experts brought in to run it. It is financed from abroad, but *local* labourers are used, who must work hard at boring tasks for a small wage, while the factory spews out wastes night and day. Who profits from this? Does this type of development reduce the poverty of the worker or not? How much does it help to bring about social and economic equality? Does it contribute to the peace and well-being of society as a whole and to the conservation of the natural environment? Of what value is it, apart from the fact that the foreign investors, along with a few local investors who usually have political interests at stake, accumulate wealth, while oppressing their own countrymen and obstructing them from having economic and political power on, or near a level with their own? The tourist industry and the hotels should ask themselves these same questions. But to even ask these questions is considered to be obstructive to development and progress. For it is difficult to measure how much so-called progress might be an impediment to a true human society, or might destroy the old culture and the quality of life. Economists and politicians usually give the matter no thought, but they continually mouth phrases about how concerned they are.

It may seem that I am not being fair to the economists, although I have taught philosophy in the Faculty of Economics at Thammasat University in Bangkok, but I would like here to show the weakness of some basic theories of that school of economists which emphasizes 'development' and which measures results in terms of mathematical figures having to do with the gross national product, the gross domestic product, and the per capita income.

When economists or members of any National Economic Development Board announce that the Gross National Product (GNP) has increased by 8% they are quoting a figure which amounts to taking the year's production and services' increase of 10% and subtracting the year's population increase of 2%. If the increase holds steady over a few years, then the economists will say the country has reached self-sustaining growth. This method of figuring is used as the basis of

Rostow's theory in regard to 'take-off', and is used in development planning.

The question we need to ask is whether it is legitimate to average this increase out in terms of per capita income, as is done. For instance, when the GNP increases by 8% it may be that 80% of the increase goes to only 10% of the population, while the other 90% of the population divides up the remaining 20%. This is what usually happens in the developing countries, so that the rich get richer and the poor poorer. But economists of this school maintain that a better division cannot yet be made, that it is not yet possible to take justice or economic equality into consideration. If we did, there would be no economic progress, and self-sustaining economic growth or the take-off point would not be reached. So it can be seen that in such countries as Thailand, the Philippines, Korea, Malaysia, and Singapore, equal opportunity is lacking to a great extent. While a few are at ease, the majority are poor, because the economists argue that for fast progress to be made it is necessary to keep the labourers and the majority of the people down, in order that foreigners and capitalists will be willing to invest. For they must keep costs low and make a fast profit if they are to continue to invest. When this happens the gross national product will increase.

This type of thinking is inescapably related to capitalistic thought which holds that the more goods increase, with the least interference, the greater the progress. Capitalistic business has only this as its goal. The goal is to make the highest profit possible by keeping the money in the hands of a few, in order to make larger investments, in order to produce more things. So production and consumption are emphasized. Advertising becomes important in order to increase greediness, and usually lust is a part of the advertising also, since greed and lust usually strengthen each other. So almost every type of advertising, in addition to luring people to want unnecessary things, will include sensual pictures.

This kind of development is always accompanied by some kind of incentive to make people want, to thirst, to desire; and the desire is usually for worldly or material things. If people were temperate in their desires, being satisfied in the material sense with the four Buddhist requisites, or a little more which might be necessary for life, in order to give them the opportunity to develop their potentialities – with each one wanting to help the other as was typical of our Buddhist village life in former times, then capitalism would fail. Is not this the reason that Western imperialism, which went to Asia along with capitalism, looks down upon the former Asian way of life as backward and uncivilized – because that kind of life provides no incentives to buy foreign goods, nor is it anxious to exploit natural resources to the maximum?

In the light of this, is it putting it too strongly to say that it is impossible for anyone who loves a peaceful life and contentment, and has an honest desire to help others, in other words anyone who takes his religion seriously, to approve national development on capitalistic lines?

By taking one's religion seriously I mean, by getting down to the heart of religion, not simply being strict about non-essentials and presenting a false front, like those Buddhists who will not slap a mosquito but have no compunction about loaning money out at exorbitant interest, or those Christians who go to church every Sunday while they use crooked business methods, even taking excessive profit from the sale of grain to feed orphans, or those Moslems who refuse to eat pork but defile themselves by devious dealing and lasciviousness, and give support to politicians who cheat the people by various devices. Is it not because of hypocrites such as these that young people turn away from religion? As one Hindu said, his father was very dishonest, but that was all right, since he was very religious – making pilgrimages, praying, and bathing in the Ganges to wash his sins away.

Religious truth is the pinnacle of religion, but it must have morality or decency as its base. Otherwise, religion is indeed an opiate, or a mask for evil people to use in reaping profits from those who are more stupid than they are.

When an economic system, based on a capitalistic market economy, requires increasing greed, both in the producer and in the consumer, can any religion encourage it?

The producer must find a way to invest his money that will bring him the greatest return, without regard to the disappearance of natural resources, or to the fact that he may be producing luxury goods. While 80–90% of the people in the world do not have the basic necessities of life, if it is possible to realize a large profit by using natural resources to produce luxuries for 1% of the people, this type of developer is only too happy to do it. When one part of the world has an overabundance of corn or wheat, they gladly burn it to raise the price even higher. Especially in developing countries, one can see the emphasis given to the hotel industry, tourism, and places catering to the physical pleasures of man, rather than to agriculture or the production of the basic requirements of food, clothing and medicine, because the more advanced nations maintain control of the markets.

The capitalistic system aims for profit, not for the general welfare of the public. Capitalists may indulge in some philanthropy, as long as the majority of the people remain under their control. But since profit is their goal, they must take every advantage they can, beginning with taking advantage of the workers and finally taking advantage of the consumers. The extent to which advantage is taken varies. In countries where labour unions are strong, where the government officials are fairly honest and efficient, and where the consumers keep up with the producers by their own organizations to test quality against price, the producers must have a higher ethical standard. In countries where these conditions do not exist, the proportion of dishonesty is higher.

Consumers ordinarily have modest desires unless they are especially stimulated to want and to buy. But in a capitalistic system the mass media is used to stimulate desire for things that are not really needed. Consumers are forced to choose between brands which in fact may be identical. The consumer becomes a victim of advertising. The claim that capitalism gives freedom to the people, by providing freedom of choice, is less than true. For advertising companies determine in large part what is to be sold, and they together with the multinational corporations deceive the people in such a way that they scarcely realize they are being duped. This is not real freedom. In countries under a dictatorship, at least the people know that the government is propagandizing them, for the propaganda is usually quite crude. But deception is more difficult to perceive when it is based on the encouragement of greed, either by using fancy decoration or by advertising methods. In such countries, wherever electricity is brought, no matter how poor a family is, it feels it must buy a television set. Television plays a most important part in the deception of the public, since it is held up as a status symbol. It is pictured as improving the way of life, so that people will sell their land if necessary to buy a television. This is but one example of tricks to induce people to buy; time payments is another.

People seem to measure development in terms of their own personal increase in new and expensive goods.

Because greed is the boss in this type of development, it causes an increase in cross-purposes and conflict; people increasingly take advantage of and oppress one another; cheating, swindling and crooked tricks inevitably follow. This is the truly materialistic way of life. Yet many nations admit with a rueful smile that is also the way of development.

It is true there are ways to correct the bad image of capitalism to some extent, at least since the time when Maynard Keynes of England attempted to improve its image by injecting some socialism. He tried to reduce the lust for things and to give some purpose beyond profit to the capitalists, so they would use at least part of their profits for the benefit of society. But capitalism was still capitalism. Even though the capitalists helped their neighbours to some extent, their basic purpose was selfish, and they only tried to disguise the fact a little more than formerly. Their way of life still emphasized physical pleasure and prestige, only not quite so obviously as before.

Can this way of life be accepted by one who turns for refuge to the Buddha? Whether or not we accept Weber's theory, it is a sad fact of history that capitalism grew and prospered first in Christian society. And the more capitalism flourished, the further society separated itself from Jesus Christ. Today capitalistic society has begun to flourish in Buddhist society. Though our Thai forebears were forced to accept it, we, or at least those of our number who now hold the reins of power in Thailand, have gladly accepted capitalism, even to the point where they have laid the plans for development along capitalistic lines, and it is now a question as to whether or not they are any longer Buddhist, regardless of what they say.

What is worse is that the majority of religious leaders, whether sons of Buddha or followers of Christ, have flocked into the capitalistic camp.

We should note plainly at this point, that it is not only the capitalistic system that has increasing the gross national product as its goal of development. The socialistic countries of Eastern Europe use the same standard as capitalistic societies in measuring the results of development. And why shouldn't they? Their oldest sister, the Soviet Union, took the lead.

In terms of human values, countries that take their refuge in materialism, no matter to what ideological camp they belong, scarcely differ in their end results. Eric Fromm has said, 'The Russians think of themselves as the representatives of socialism, because they use Marxist ideological terminology, not realizing how closely their system resembles the fully developed capitalistic system. We in the Western camp, on the other hand, believe we are the representatives of an individualism which encourages each person to take the initiative, with a humanitarian ethical system; thinking that these are our ideals. But we fail to see that in truth many of our own social institutions closely resemble those of the communism we detest.'

Development which has quantity as its basic goal has many weaknesses as we have pointed out. It is not that economists are not aware of the weaknesses. As Keynes once attempted to correct the weaknesses of capitalism, so more recently the United Nations Social Development Research Institute has sought ways by which income might be more nearly equalized throughout society. It is an admirable effort and should be encouraged. But to date it has failed.

The Institute mentioned above has listed about seventy-seven items which can be assumed to be important in development, including such things as long life, sufficient vitamins in the diet, the number of pupils enrolled in school, the number of people who use electricity, the numerical ratio of buildings to people, the opportunity of access to radios and television, the amount of agricultural produce in relation to the number of male farm workers, the use of currency in exchange for food, etc. It is held that the more improvement and expansion in these items, the greater the development. This means that the Institute is attempting to turn the members of the United Nations away from their preoccupation with increasing the national income as the goal of development, and to lead them to make a just and equitable distribution of the income.

This is a real step forward, yet it still does not go beyond the usual quantitative definition of development. Measurements are still made in terms of quantity. None of the seventy-seven items can be used to measure life qualitatively. And this method does not take into consideration the means used in increasing production, but simply assumes that the more production the better, as long as there is a just distribution.

So we can say that the error in development lies in making quantity its goal, in continually trying to measure results in terms of materialism and 'modernity'. This type of thinking has its roots in Western Europe at the time of the industrial revolution. But the question is not asked as to whether this way of life is really desirable or not. Rather the social order of the Westerner is simply accepted as the goal, as the model to imitate, and the only question asked is how many years before a country like Thailand will catch up with England or Japan or Russia or the United States?

I am glad to say that leading thinkers including economists now see more clearly that this type of development solves no problems. The United States, which once was the land of dreams of the free world, is full of social injustice and poverty within, and takes unfair advantage of countries outside its boundaries. It has food to throw away, and with only 6% of the world's population, it consumes almost 50% of the world's irreplaceable natural resources. At the same time, the rich become drug addicts, crime increases, and the old Christian values are given up. This seems to be happening all over the country that once was held up as the prime example of this type of development.

It is evident that our world is caught up in a cycle. Especially in quantitative development, the further it goes the more problems appear, faster than they can be solved, and the technocrats are not able to stop the spiralling because (1) they are afraid that if the quantity is not increased everything will come to a standstill, or all the systems will go haywire leading to possible ruin. For instance, the population will increase and there will be insufficient food, leading to clashes. Actually in this regard there would be enough food, if it were distributed equitably and used without waste. The problem is that those who have the surplus refuse to share it, because (2) they want to maintain the status of the rich. Their hope is that by increasing production most of the poor will receive a portion of the increase, continually raising their standards. But our nature is such that once we ourselves have become more comfortably situated, even though we see some injustices appearing, we don't get too excited about them if they do not touch us too much. Besides, if we do something about them, we might get hurt.

Is it not for these reasons that development has worked out in such a way that the gap has grown between the rich and the poor, and between the wealthy nations and the poor nations? In Thailand, since development planning began, a few wealthy people in Bangkok have become continually wealthier, while the people of the northeast have become poorer, not to mention conditions in other parts of the country. And up until now, there has been no indication (including the present dictatorial regime which is a swing from the former democracy) that my country is considering a change in its development policies, but it goes blithely on following the blueprints of the capitalistic economists. If anyone raises objections to these methods or this type of thinking, he is labelled a rabble-rouser, a proponent of communism, or else he is accused of disloyalty to the Nation, Religion and the King. Is it not time that we should speak the truth, and especially those who hold themselves to be religious? We must be honest, and if we are honest, we must admit that this type of development has not added to the happiness of the people in any real human sense, but on the contrary has taken a form that to a greater or lesser extent is permeated throughout with crooked deceptions. Moreover, we must not forget that the increase in production through the use of modern machinery to exploit natural resources cannot go on forever. Oil, coal, and iron, once they are gone, cannot be brought back. As for the forests and some wild animals, when they are depleted, if they are to be brought back, it will not be in our time or our children's time. Production on a grand scale not only uses up the raw materials, it also destroys the environment, poisoning the air and the water, the fish and the fields, so that people are forced to ingest poison continually. But we do not need to expand further on how man takes advantage of his fellow man.

In short, any country that feels itself so inferior as to call itself developing or underdeveloped, cannot and should not try to raise itself up through this kind of quantitative development in order to put itself on a par with those nations which brag that they are developed.

It cannot do so because, as Everett Reimer has said, if every country were like the United States, 'the oil consumption would be increased fifty times, iron one hundred times, and other metals two hundred times. And the United States itself would have to triple its use of these materials simply in the process of production itself.' There are not enough raw materials in the world to do this, nor would the atmosphere be able to take the change. The world as we know it would come to an end.

You will, no doubt, surmise that I do not agree with that form of development which aims at quantity, and not even that form of development which has as its objective the improvement of the quality of human life, yet still stresses material things. In reality, the latter too diminishes the quality of human life.

It is not only that materialism fosters violence, but modern applied science also destroys the values of time and space. To a materialistic civilization, time means only that which a clock can measure in terms of workdays, work-hours, work-minutes. Space simply has three dimensions which are filled with material things. That is why Buddhadāsa says, development means confusion, for it assumes the more the merrier. The longer one's life the better, with no thought of measuring the real value of a long evil life as against that of a short good life. This is contrary to the teaching of the Buddha, who said, the life of a good man, however short it may be, is more valuable than that of an evil one, however long he lives.

As a matter of fact, it is only religion, which puts material things in second place and keeps the ultimate goals of development in sight, that can bring out the true value in human development. For even in the matter of judging the value of development, from the point of view of ethics and morality, it is difficult to keep material consideration from being the sole criteria.

From the Buddhist point of view, development must aim at the reduction of craving, the avoidance of violence, and the development of the spirit rather than of material things. As each individual progresses, he increasingly helps others without waiting for the millennium, or for the ideal socialist society. Co-operation is better than competition, whether of the capitalist variety which favours the capitalist, or the socialist variety which favours the labourer.

From the standpoint of religion, the goal can be attained by stages, as evil desires are overcome. So goals are perceived in two ways. From the worldly standpoint, the more desires are increased or satisfied the further development can proceed. From the religious standpoint, the more desires can be reduced the further development can proceed.

Western civilization erodes Christianity, or at least real Christian spiritual values, becomes merely capitalistic or socialistic, and aims to increase material goods in order to satisfy craving. The capitalist variety wants to raise the material standard of living of other groups, if possible, providing the capitalists themselves can stay on top. The socialist variety reverses it and wants the majority, or those who act in the name of the majority to oppress the minority or those who are opposed to them.

The value scale of Western-type development emphasizes extremes. The richer the better; the capitalists apply this to the wealthy, and the socialists to the labourer. The quicker the better. The bigger the better. The more knowledge the better. Buddhism, on the other hand, emphasizes the *middle way* between extremes, a moderation which strikes a balance appropriate to the balance of nature itself. Knowledge must be a complete knowledge of nature, in order to be wisdom; otherwise, knowledge is ignorance. Partial knowledge leads to delusion, and encourages the growth of greed and hate. These are the roots of evil that lead to ruin. The remedy is the three-fold way of self-knowledge, leading to the right speech and action and right relations to other people and things (morality), consideration of the inner truth of one's own spirit and of nature (meditation), leading finally to enlightenment or complete knowledge (wisdom). It is an awakening, and a complete awareness of the world.

When one understands this, one understands the three characteristics of all things from the Buddhist point of view: their unsatisfactoriness, their impermanence, and their lack of a permanent selfhood.

True development will arrange for the rhythm of life and movement to be in accordance with the facts, while maintaining an awareness that man is but a part of the universe, and that ways must be found to integrate mankind with the laws of nature. There must be no boasting, no proud self-centred attempts to master nature, no emphasis placed on the creation of material things to the point where people become slaves to things and have no time left for themselves to search after the truth which is out beyond the realm of material things.

In 1929, Max Scheler formulated a remark which is just as true today as then. He said,

> We have never before seriously faced the question whether the entire development of Western civilization, that one-sided and over-active process of expansion outward, might not ultimately be an attempt using unsuitable means – if we lose sight of the complementary art of inner self-control over our entire underdeveloped and otherwise involuntary psychological life, an art of meditation, search of soul, and forebearance. We must learn anew to envisage the great, invisible solidarity of all living beings in universal life, of all minds in the eternal spirit – and at the same time the mutual solidarity of the world process and the destiny of its supreme principle, and we must not just accept this world unity as mere doctrine, but practice and promote it in our inner and outer lives.

This is indeed the spirit of Buddhist development, where the inner strength must be cultivated first; then compassion and loving-kindness to others become possible. Work and play would be interchangable. There is no need to regard work as something which has to be done, has to be bargained for, in order to get more wages or in order to get more leisure time. Work ethics would be not to get ahead of others but to enjoy one's work and to work in harmony with others. Materially there may not be too much to boast about, but the simple life ought to be comfortable enough, and simple food is less harmful to the body and mind. Besides, simple diet could be produced without exploiting nature, and one would then not need to keep animals merely for the sake of man's food.

In *Small Is Beautiful*, E.F. Schumacher reminds us that Western economists go for maximization of development goals in a material sense so that they hardly care for people. He suggests Buddhist economics as a study of economics where people matter. He says that in the Buddhist concept of development, we should avoid gigantism, especially of machines which tend to control rather than to serve man. With gigantism, men are driven by an excessive greed in violating and raping nature. If these two extremes (bigness and greed) could be avoided, the Middle Path of Buddhist development could be achieved, i.e. both the worlds of industry and agriculture could be converted into a meaningful habitat for man.

I agree with Schumacher that small is beautiful in the Buddhist concept of development, but what he did not stress is that cultivation must first come from *within*. In the Sinhalese experience, the Sarvodaya Shramadana movement applies Buddhism to the individual first. Through cultivated individuals a village is developed, then several villages, leading to the nation and the world.

The guideline for the movement is the use of the Four Sublime Abodes (*Brahma Vihāra*) to develop each individual. The steps to be taken are as follows:

1 *Nettā*: Loving kindness towards oneself and others. We all desire happiness. We should try to be happy. Through the precepts and meditation, a happiness state could be created. The mind will feel amity and harmony with oneself as with others. It renders assistance and benefits without ill-will, without the malice of anger and of competition. Once one is tranquil and happy, this tranquillity and happiness could spread to others as well.
2 *Karunā*: Compassion can only be cultivated when one recognizes the suffering of others and wants to bring that suffering to an end. A rich man who does not care for the miserable conditions of the poor lacks this quality. It is difficult for him to develop himself, to be a better man. Those who shut themselves in ivory towers in the midst of an unjust world cannot be called compassionate. In Mahāyāna Buddhism, one should vow to become a Bodhisattva who will forego his own nirvāna until all sentient beings are free from suffering. So one should not remain indifferent, but must endeavour to assist others to alleviate their suffering as much as one can.
3 *Nuditā*: Sympathetic Joy is a condition of the mind which rejoices when others are happy or successful in any number of ways. One feels this without envy, especially when a competitor is getting ahead.
4 *Upekkha*: Equanimity means the mind is cultivated until it becomes evenly balanced. It becomes neutral. Whether one faces success or failure, whether one is confronted with prosperity or adversity, for oneself or for others, one is not moved by it. Whatever one cannot do to help others, one is not disturbed about it (having tried one's best).

The Four Sublime Abodes should be developed step by step from the first to the last. Even when one is not perfect, one must set one's mind toward this goal, otherwise one's dealing with others will tend to be harmful – to oneself and to others – one way or the other.

Having developed oneself toward happiness and tranquillity rather than toward worldly success and material progress, then a Buddhist is in a position to develop his community, starting with his family and his village. He must first be awake before he can awake others. An individual who is awake is called *Purisodaya*. By sharing his awakening with others, the whole village could become awake – *Gamodaya*. Once several villages awake, the whole nation could perhaps be saved from materialism and Western economic development models. In the case of Sri Lanka and Burma, when proper efforts are put along these lines, the chance is greater than in the case of Siam, which lost so much confidence in the Buddhist heritage (especially among the ruling élites). The Western developmental process has been accelerated far too much during the last decades. But, of course, if Thailand could free herself from being a junior ally of the USA, and also be saved from Communist aggression, she too could have a good chance to opt for the Middle Path of development and become a *Desodaya*, an awakening nation. The hope for several nations to awake so that the world could be saved and we would reach the state of universal awakening for all – *Sarvodaya* – at this stage is very remote. Yet, we must not despair, and we must live in hope and practice what we can.

In the Buddhist experience of Sri Lanka, the driving force to develop from the village level upwards comes from the Buddha's teaching of the Four Wheels. As a cart moves steadily on four wheels, likewise human development should rest on the four *dhamnas*, namely, Sharing, Pleasant Speech, Constructive Action, and Equality.

1 One must share (*dāna*) what one has with others – be it goods, money, knowledge, time, labour, or what have you. This is still practiced in most village cultures. Sharing does not mean giving away what one does not want, or giving in order to get more. We should strengthen the Buddhist concept of *dāna* practiced in the villages and spread it to counteract the invasion of materialism and the new value system of competition – by sharing, by giving freely rather than by buying and selling. In Sri Lanka, they share labour, with Buddhist ceremonies in the background, as the Thais still do in remote villages where they find such fun in work.
2 Pleasant Speech (*piyavāca*) not only means polite talk, but means speaking truthfully and sincerely, regarding everyone as equal. This too is strong in village culture, although villages have received glamourous propaganda treatment by politicians and advertisers, who are full of deceit and make one buy things one does not really need, or make one hope for something which is not really possible.
3 Constructive Action (*atthacariya*) means working for each other's benefit. Here Schumacher's recommendation for intermediate technology and the proper use of land would be relevant.
4 Equality (*samanatata*) means that Buddhism does not recognize classes or castes, does not encourage one group to exploit the other. So Buddhist socialism is possible, without state capitalism or any form of totalitarianism.

The development toward Buddhist socialism means that equality, love, freedom, and liberation would be the goal. A Buddhist community – be it a village or a nation – would work for harmony and for awakening, by getting rid of selfishness of any kind – be it greed, hatred or delusion. Such a development would entail truth, beauty, and goodness – be it big or small.

13

The Idea of Development. The Experience of Modern Psychology as a Cautionary Tale and as an Allegory

Ashis Nandy

I

The modern idea of development, everyone knows, began as the idea of economic growth. The roots of the idea may have been in the Judaeo-Christian worldview, but the idea emerged, as Ivan Illich points out, as an empirical as well as a normative category in the 1940s when, of all persons, President Harry S. Truman first used it in its present sense.[1] Like most important concepts in the social sciences, the idea of development, too, has kept on expanding in scope during the four decades it has been in currency. Today, efforts are still being made to enlarge the concept of development and include within it larger chunks of social reality, so that development becomes something more than economic growth and something more than the removal of economic and non-economic impediments to economic growth. There is now more stress on social and political development, on social justice and elimination of institutionalized violence, on scientific and technological growth and, more recently, on human development including the various modalities of consciousness-raising and even aspects of self-actualizing. Even within the objectivist schools there is now greater concern with subjective factors such as 'internal repression', consumerism, instrumental reasoning and, of course, alienation. Development has now become a proper ideology and a worldview. It has now something to say on nearly everything.

Simultaneously, over the decades the idea of development has become partly independent of the justifications provided for it. It itself now justifies structures and processes. Like modern science and technology, development too has become a new reason of State in many societies. Today, hundreds of thousands of citizens can be legitimately killed or maimed or jailed in the name of development. And hundreds of serious scholars all over the world would be perfectly willing to provide serious scholarly arguments for turning the other way when such organized, planned oppression takes place. The idea of development now is bonded to a vision of a future society which demands sacrifices – often blood-sacrifices – at a scale which would put to shame any tribal society.

This expanding scope, meaning and power of the idea of development prompts me to examine here the experience of contemporary psychology, the one field of human knowledge where the idea of development has been central for many decades, and to see what lessons the discipline has to offer to the students of human development in general. After all, both modern psychology (with its concern with child development, personality development and cognitive development) and traditional psychology (with its concern with individual self-realization, maturity and personal growth through cycles of self-training and self-development) have faced some of the same choices which the student of social development faces today. These choices should have something to say about the problem of development, even if in a language which is esoteric from the point of view of the modern man.

The main lessons I would draw from the following discussion will, I suspect, be nebulous. But I hope they will not be trivial, not at least to those interested in the encounter between the worldview of development and the ancient civilisations which do not believe in development and yet have to cope with it. However, to the extent that such civilisations classify differently the problems now handled under the rubric of development, to the extent that even an 'emic' or inside-out approach to 'development' is to these civilisations an ideological imposition, these lessons will, I suspect, look relevant to only those in the modern sector, in the West as well as in the East.

II

Over the last eighty-five years, a number of attempts have been made to define the fully 'developed' person under the rubric of the psychology of the creative, self-actualizing or healthy person. There have been standard clinical views on the subject, humanistic critiques of such views, and, even, as in Sigmund Koch, critiques of such critiques.[2] Instead of venturing an alternative differentia of psychological development (which would help retain the developmental framework while changing the specifics within it) and instead of summing up the Indian experience with the concept of development (which will make sense mainly as a crosscultural data set), I should like to begin with the old debate in modern psychology over the definitions of health and illhealth.

Here on this issue, on one side there are those who see the normal and the abnormal as clearly disjunctive; on the other, there are those who see the two as the ends of a continuum. Both approaches derive sanction from the history of modern medicine; both enjoy a certain legitimacy within the Western intellectual tradition; and by now both have developed some identifiable features. To the first group belong most biologically-minded psychologists and most behaviourists. To the second belong most 'depth' psychologists, including the Freudians, the existentialists, the humanistic psychologists and the structuralists. That these are not clear-cut schools is evident from the fact that Gordon Allport, a humanistic psychologist, clearly falls into the first category with his concept of functional autonomy of personality processes; and psychoanalytic ego psychology, with its stress on the conflict-free ego sphere, comes close to the position that health is at least partly an autonomous process.[3]

Both the positions – the one which sees a clear break between mental health and illhealth and the one which sees a continuity or a non-orthogonal relationship between the two states – have sometimes yielded major insights and humane interpretations of persons and societies; both have sometimes made psychologists wear intellectual and moral blinkers. The first position, however, has been more enthusiastically blessed by the 'toughminded' psychologists, the second by the 'woollier' lot. The second position is also generally more compatible with many primitive or traditional psychologies which see madness as an altered state of consciousness and as an opportunity to experiment in transcendence and self-exploration.

The crucial difference between the positions however is the following: According to the first, health is a matter of a personality development which follows a 'natural' chronological line and certain population norms. Illhealth is a deviation from these norms and, thus, it needs to be corrected by a dimensional or sectoral intervention in the personality, that is by 'managing' a particular pathology – a trait or a behaviour or a need or a symptom or even a complex or a neurosis. A crude caricature of the approach based on the idea of discontinuity would be to set up statistically a series of indicators to separate the healthy from the unhealthy and work with a concept of mental hygiene which would aim to remove, one by one, all signs of pathology from a person to make him fully healthy.

According to the second position, health is only a more humane or creative use of illhealth; and illhealth, in turn, is an 'inferior' or less creative dynamic of health. It follows that both health and illhealth are holistic states and 'organisational principles' rather than sums of discrete elements or sums of scores on a list of indicators. A good instance of the second position is those studies of creativity which show that creative persons score high on most indicators of illhealth and yet are creative by virtue of their greater ego strength. This strength creates new resources for creativity precisely out of those elements of personality which are, statistically speaking, pathological.

III

The debate on the continuity between health and illhealth – and by implication that between the psychologically developed and the underdeveloped – has gone on for a long time. We now know the basic arguments of both sides. However, underlying this debate has been another latent, second-order debate among those who believe in the continuity between health and illhealth. The nature of this other debate can be summed up in the form of a question: should one see the continuity from the illhealth end (that is, pathographically) or from the health end (that is, non-pathographically)?

In the first instance, the continuity between health and illhealth is established when health is shown to be a superstructure built on a base of illhealth – Sigmund Freud's *Leonardo da Vinci* is as good an example as any.[4] Into this category fall much of orthodox psychoanalysis, much of radical psychology and much of what is called psychohistory.

In the second instance, the continuity between health and illhealth is established when illhealth is shown to be a variant of health or as a superstructure built on basic healthiness – Gregory Bateson's and R.D. Laing's work on schizophrenia immediately comes to one's mind.[5] Within modern psychology such instances are rather rare. In this category, however, fall many non-Western idea systems which ascribe mystical or sacred meanings to mental illness; some traditional Western schools of thought which see madness as a possibility, if not as another form of normality; and the inarticulate philosophies of life of the highly creative, deviant individuals in the modern West who have gone through the proverbial journey through madness (not to transcend madness but to use madness as a window to an alternative world). Vincent Van Gogh and Fyodor Dostoevsky are obvious examples.

The modern age has made its choice. It rejects – or banishes from the world of science, rationality and pragmatism – visions which see mental illness as an opportunity and as a possible window to the infinite, the unknown and the invisible. Modernity may sometimes see health and illhealth as the two ends of a continuum but its basic goal is to eliminate illhealth constantly and completely. Hence, the evolutionism which colours the modern understanding of both personal life cycles and cultural differences. Hence, the set of relationships which pervades so much of modern thinking on mental health:

scientific: irrational: historical: mythical::
adult: infantile:: civilized: savage:: modern: primitive:: sane: insane:: developed: undeveloped

Though there have been some like Emile Durkheim who never lost sight of the dialectic between – and mutual interdependence of – health and illhealth, the modern consciousness has mostly tried to attain its critical edge by unmasking health, to reveal underneath the hidden illhealth. Also, after identifying the pathology of normality, modern critical consciousness has to try to defeat and eliminate that pathology to restore true normality. The aim is to first detect the irrationality and then destroy its basis so as to reinstate full rationality. This is the other side of Sigmund Freud's hope that psychoanalysis would help the ego to gradually supplant the id.

Such a format of demystification was bound to be cannibalized sooner or later by everyday life and 'normal' science. The idea of normality – and the concept of rationality on which that normality bases itself – was easy to co-opt in a world increasingly becoming managerial and techniques-oriented. Once the idea of normality was operationalized in a formal or mechanical way (that is, as a uni-directional exercise in unmasking), it became easy to fit it in within the dominant consciousness as a legitimization of that consciousness.

Here lies the politics of the anti-psychiatry movement and that of the related studies of madness in the Western civilization. Michel Foucault, R.D. Laing and Thomas Szasz – all of them recognize the possibilities, political and/or visionary, of at least some forms of mental illness.[6] All of them refuse to take for granted the given definitions of health and, what I have called, a uni-directional format of unmasking. To the extent that these writers subscribe to the view that a normal person or society is not that normal after all, they are influenced by modern Western critical-scientific traditions. But to the extent that they describe the way societies construe insanity out of alternative or less popular modes of consciousness, and to the extent that they believe that the abnormal is not that abnormal after all, they are the rediscoverers of the non-pathographic approach within modern psychology.

I hope my tone does not suggest that I deny the significance of criticisms of 'normality'. I grant that the debunking of health can be an excellent baseline for assessing the pathology of normality, conventional rationality and everyday life. Such debunking does offer a powerful social criticism of the banality of everyday life and the excesses of, what Kuhn calls, normal science.[7] I reject also the popular criticisms of depth psychology which blame the discipline for its 'over-dependence' on data from the clinic. I consider such dependence to be an oblique expression of what was once a serious problem, namely the refusal of academic psychology to admit the continuity between health and illhealth. If you accept such continuity, illhealth can as easily be a datum for the psychology of health as health can become a datum for the psychology of illhealth.

However, I believe that our times have also revealed the clay feet of the great tradition of critical social analysis built by Vico, Nietzsche, Marx and Freud, who have updated the style of criticism Galilean science pioneered in post-medieval Europe.

First, each of the main schools of thought in this tradition has developed a stratum of experts. The scientist, the revolutionary vanguard, the conscientizer, the psychoanalyst, the child psychologist guiding child development – they may or may not constitute an economic élite but they surely are power élites extracting substantial psychosocial surpluses. Willy-nilly they not merely hegemonize all cultural discourses but also define the discourse of their victims. The doctor today is not only an expert; he has marginalized the rights and even the speech of his patient on the patient's body. The idea of a vanguard today not only allows one to claim monopoly on the understanding of a revolution but also on defining the critics of the revolution, internal and/or external.

Second, most critical theories tend to totalize. Not only are they afraid of multiple interpretations of the same phenomenon, they consider such multiplicity unscientific, irrational and unaesthetic. So much so that all alternative theories are seen by each critical theory as vested interests and pseudo-alternatives. Such a perspective is maintained with the help of a certain blindness, in each such theory, towards the full implications of its own model. Thus, there is Freudian or Marxist demystification of almost everything except Marxism and Freudian thought themselves.

Third, each critical tradition has set important limits to its interpretation. Marxism does not allow its adherants to look beyond the 'base'; psychoanalysis refuses to go beyond psycho-sexuality. As a result, in Marxism, new forms of oppression released by the revolutionary process are laundered through the subtheory of counter-revolution. In psychoanalysis, the inner contradictions of the 'successfully' analyzed patient definitionally becomes a matter of non-antagonistic contradictions.

All said, the critical tradition may have liberated others but it has failed to liberate its own votaries. Indeed, it has shown a fearsome ability to bind its allegiants into a lifelong psychological serfdom. There is no protection within the tradition for the future generations, for the savages and for the victims of history wanting to defy the existing models of liberation.

IV

For the moderns, criticism has to have a scientific baseline which closely defines the truth and, therefore, defines even the validity of alternative baselines. For at least some savages, criticism is not so much a search for truth as a dialogue among different constructions of truth – among cultures, levels of living and states of consciousness. To these savages, the continuity between the normal and the abnormal is a creative dialectic between two views: one which sees all normality as a special case of abnormality (as in orthodox psychoanalysis) and the other which sees all abnormality as a special case of normality (as in many traditional visions of man). The contemporary world stresses the first part of the story and ignores the second. When it accepts the second, it rejects the first as an attitude which is irreverent (as do the modern pedlars of instant-salvation from the East trying to speak to the Western counter-cultures or alternative lifestyles). The choice in modern times boils down to one between the view which has reached its apogee in T.H. Huxley's or, for that matter, Joseph Stalin's concept of scientific rationality and the politically less powerful view of the traditionalists co-opted or tamed by the modern world – the apolitical gurus operating within the modern sector – from Reverend Billy Graham to Orient's own Acharya Rajneesh.

Perhaps within the clinic, it matters less what one's therapeutic ideology is. Suffering as a direct experience has some ability to shape the diagnosis and the therapy of the more sensitive therapist. In everyday therapy, a simple approach (whether exclusively pathographic or exclusively non-pathographic) may even be an advantage for the therapist, provided his antennae are tuned to the nuances of human suffering. Such simplicity, however, becomes a disaster when one uses depth psychology for evaluating cultures or for serving as the baseline of a normative social psychology (the use to which the late Erich Fromm wanted to put psychoanalytic social psychology when he pleaded for more psychology of the unconscious, and devalued ego psychology as the cat's paw of conformism and *status quo*).[8] It is one form of social criticism to show that underlying our concept of health are major personal or cultural psychopathologies deriving from everyday life and from a given period or culture. It is another form of social criticism to show that what the powerful and the rich define as an abnormality is but another form of health, opening up another set of possibilities. It is the second form of criticism which has been ignored in modern psychology.

This may sound like only a more complicated plea for cultural relativism or for humanistic psychology. But the approach has a scope larger than that. It has a built-in criticism of the worldview of development and its evolutionist trappings within psychology. Certainly, the approach has resisted the entry of the theory of progress into the ethnopsychologies of many cultures. The Indic civilization, for instance, has traditionally treated almost all the states of consciousness, including the various forms of meditation, trance and vision, as 'the substance and occupation of the highest life' and as the 'foretaste of a higher existence'. The civilization as a whole has never tried to denounce these states as morbid, sick, infantile or primitive or as objects of study by the scientifically minded.

Even though I have pleaded for a dialectic between the pathographic and non-pathographic views, I believe that the non-pathographic view must be particularly defended and nurtured in our times. For it is now the 'minority' view in the accessible world of knowledge. I am personally uncomfortable with a purely non-pathographic view but I also know that it serves as a language in which at least some victims of history articulate their understanding of man-made suffering. And the challenge of our times is to restore, at least partly, the categories of such victims in our theories of oppression.

Also, the non-pathographic view defends the self-esteem of societies which have been studied only in terms of categories produced by the psychological needs of the modern civilization and by the psychopathologies such a civilization fears. Such a defence counters the pathographic view which has often reduced entire societies and cultures to the status of deviant or abnormal cases or to the status of primitive or infantile stages of history. The refusal to give in to a pathographic view not only allows one to conceptualize each mix of normality and abnormality as a creative possibility – in the process pluralizing the concept of normality itself – it also sees these alternative forms of normality as vantage points from which conventional uni-linear views of normality, health and personality development can be criticized. Thus (1) it legitimizes external, in addition to internal, criticism of modernity and enlarges the scope for basic criticism of societies, including the modern ones: (2) it makes each modal personality type or culture simultaneously a subject and an object, thus de-hierarchizing and de-expertizing the world of conventional crosscultural psychology; and (3) it denies intrinsic, permanent legitimacy to any worldview, any science of change, and to any theory of psychological development. It thus protects the nonconventional concepts of health against a hegemonistic vision of the ideal person and against a monopolistic eupsychia which has the political and economic backing of the dominant cultures of the world.

The third point is especially important. The cultural relativists may find such negativism even more difficult to swallow than the universalism in which much of conventional psychology is mired. But such critical or negative relativism, unlike uncritical cultural relativism, refuses to grant absolute legitimacy to any state of consciousness. Each consciousness, this point of view affirms, has creative and humane potentialities but each consciousness also can be – and frequently is – oppressive.

V

The first lesson of the psychologist's hobnobbing with the concept of development, thus, is rather simple. Growth, development, maturity are not innocent scientific terms. They, too, hierarchize states of consciousness and forms of living. They, too, legitimize authority the way some religions and myths did – or do – in many traditions. The idea of development especially is now part of a larger secular theory of salvation and a new reason of State. In the name of development new forces of oppression can be unleashed. Yet, unlike the oldest reason of State, national security, this new reason of State has not been subjected to systematic demystification. And this demystification, the experience of psychology suggests, cannot be brought about the way the protagonists of the idea of alternative development believe, namely by extending the meaning of development by putting in more and more variables or indicators into the measurement of health. Such an expansion of meaning only expands the area of social planning and expert intervention to new areas of a person's or group's life. If one wants to demystify the ideology of development, one must learn to go outside the modern worldview and, more specifically, outside the theory of progress.

The second lesson is indirect and it seemingly runs counter to the first. It is best expressed in the popularity of the rhetoric of development in societies like India. Psychologically, the theory of progress is no longer a Western category. It has shown the ability to tap aspects of the national character or modal personality of most non-Western cultures. Criticisms of normality (as in Fromm's concept of 'market orientation' or Riesman's 'lonely crowd') have failed to realize the pull which the consumerism of the Market and the molecular individualism of the mass society exerts on those at the margins of the world.[9] There are now persons in nearly all traditional societies who are willing to adopt the evolutionist point of view, so as to taste the bitter-sweet fruits of scientific growth, 'over-development', ecocidal mastery over nature, achievement-oriented individualism and high GNP. Modernity as a preference pattern is not a unique feature of modern societies; it is present in many non-modern cultures and persons as a latent vector. Perhaps, the humiliation and oppression which the moderns impose on the non-moderns have delegitimized some of the traditional restraints and allegiances. In the minds of many, the old restraints and loyalties now mean a non-alienating but low-consuming lifestyle which has to be lived under indignity, an austerity which has become another name for imposed manmade deprivation, and a nonviolence which seems to spring from the impotency of the weak. In a cannibalistic world of ideas, wherever the conventional idea of development co-exists with alternative visions of an ideal person, society or culture, the conventional idea has made it difficult to convey alternative ideas of social intervention to a section of the oppressed – even in the name of the grandeur of their civilizations, the beauty of their low-consuming lifestyles, or the threat of overdevelopment and hypernormality. Some of the most materialistic, individuated and competitive persons can now be found among the materially deprived and the politically dominated.

Third, the experience of psychology seems to make a modest case for a dialectic between the pathographic and the non-pathographic approaches to health, even while making clear the basic problems with the modern concept of mental health. Such a dialectic allows one to view the illhealth called underdevelopment or nondevelopment as the name of a presently unpopular, cornered but nevertheless valuable strain within the global community of cultures – indeed, as a strain which is sometimes not only a normal or healthy response to oppression and humiliation but as an intrinsically valuable trait in personality or culture. Some of the values generated by or associated with underdevelopment, poverty and non-modernity may have even become central to human survival and creativity today. Yet, the dialectical vision does not force us to jettison all criticisms of cultures.

The final lesson from development psychology is that the irrational, infantile, underdeveloped self can be not merely a 'character defence' of the weak, a part of their resistance to the oppressive and brutalizing consciousness which dominates the globe; it is a defence which defends values denied in the language of domination. To ensure this protection, the nondevelopmental consciousness has to defy even the vague evolutionism of some like Abraham Maslow who believe that the 'deficiency needs' and 'growth needs' are diachronically related. That is, the material or basic needs of the individual must be satisfied before the nonmaterial or higher-order needs can become salient in his personality. Insane though it may seem, the nondevelopmental principle affirms following ancient wisdom, that the so-called deficiency needs and growth needs can be – indeed must be – pursued synchronically because the two kinds of needs are vital to each other. Either kind, if pursued in isolation, becomes internally contradictory and self-destructive. This is another way of saying that the emphasis on nonmaterial needs in many cultures is not an idealist defence against unfulfilled biological and material needs but a recognition that some 'higher-order' needs must be met in order to meet the 'lower-order' needs.

By a twist of fate, this recognition has become a clue to global survival today. No country now, however committed to the material welfare of its citizens, can ensure that welfare unless it respects the sanctity of human life and nonhuman nature, the freedom and dignity of its citizens, and the traditions of its peoples. That is the meaning of the radical ecological movements in some of the most oppressed parts of India today. Resistance has to be today a part of a larger self-discovery. And that self-discovery has been, in many cultures, a part of a spiritual quest. Delegitimizing that quest is to delegitimize the resistance.

Let us be warned by the prosperous consumption-machine called the modern industrial worker, wearing a look of satisfaction with his material success, as if only to scandalize his radical well-wishers. His concept of success threatens to destroy the minimal interpersonal bonding and social consensus on which even the competitive individualism of modern capitalism is built. The faith in the priority of basic needs has already helped to build a pseudo-culture in which even the chance of meeting the basic needs of some (by some extracting surpluses from the others) becomes smaller and smaller. For that exploitation, too, requires a minimal legitimization of power, based on a vision – however false – of a good society, some human control over technology and knowledge, and some theory of suffering and transcendence.

Admittedly, many meditational states and forms of asceticism, when pursued with a total unconcern for the deficiency needs of the majority, can become, in the psychoanalytic sense, a defensive denial of real-life suffering. But the spirit-denying consumerism of the rich and the obsessive-compulsive concept of monetized work of the powerful, too, can be neurotic defences against recognizing the decadence and degradation of the privileged. Concretization – of happiness, success, morality, creativity and work – can be pathological, too. As this century has shown, there is not only Maslow's theory of the hierarchy of needs but also a half-articulated anti-Maslowian logic which says that if freedom can be an ego-defence and not a real substitute for bread, bread too can be an ego-defence and not a real substitute for freedom.

I leave it to minds more creative than mine to work out the new rules of interpretation of this old dialectic.

Notes

1 Ivan Illich, 'The Delinking of Peace and Development', *Gandhi Marq*, 3, 1981, pp. 257–65.
2 Sigmund Koch, 'Reflections on the State of Psychology', *Social Research*, 38, 1971, pp. 669–709.
3 Gordon Allport, *Personality: A Psychological Interpretation*, New York: Henry Holt, 1937; Heinz Hartmann, *Ego Psychology and the Problem of Adaptation*, New York: International University Press, 1939.

4 Sigmund Freud, *Leonardo da Vinci: A Study in Psycho-Sexuality*, translated by A.A. Brill, New York: Random House, 1947.
5 Gregory Bateson, *Steps to an Ecology of Mind*, New York: Ballantine, 1972; R.D. Laing, *The Divided Self: An Existential Study in Sanity and Madness*, Harmondsworth: Penguin, 1965.
6 Michel Foucault, *Madness and Civilization: A History Of Insanity in the Age of Reason*, New York: Pantheon, 1965; R.D. Laing, op. cit.; Thomas S. Szasz, *The Manufacture of Madness: A Comparative Study of The Inquisition and the Mental Health Movement*, New York: Paladin/Granada, 1973.
7 Thomas Kuhn, *The Structure of Scientific Revolution*, Chicago: University of Chicago, 1962.
8 Erich Fromm, *The Crisis of Psychoanalysis: Essays on Freud, Marx and Social Psychology*, Harmondsworth: Penguin, 1973.
9 Erich Fromm, *The Sane Society*, New York: Holt, Rinehart and Winston, 1947; David Riesman, *The Lonely Crowd*, New Haven: Hale University, 1950.
10 Abraham H. Maslow, *Toward a Psychology of Being*, New York: Van Nostrand, 1962.

Index

Achburn, S. Smith 51, 53, 62
achievement and success 5, 10, 12, 16, 35
   utilitarian 38–40, *t. 40–2*
action, collective 193, 197–203, 207, 210
activation theory 66, 67, 68
activity *see* time, use
actualization, self 14–15, 82
   potentialities *t. 52*, 76–8, 134
adaptation
   mechanisms 39, 113–15
   and reality 30, 77
   social 30, *t. 40–1, 42*
      and human development 126–38;
      failure in adaptation 131–3;
      predominance of adaptation 128–31;
      transcendence over adaptation 134–6
   syndrome 130–1
   see also change; character, social
adolescence 124
advertising 230, 237, 239
Africa 213–32
   concept of man 217–19
   culture 214–15, 217–32
   intellectual 230–2
   modern 222–28
   morality 219–21
   religions 221–2
   socialism 228–1
age
   life-stages 50–5, *t. 52–3*, 57–62
   longevity 84, 242–49
   sensorial 51, *t. 53*, 62
agency 160
   alienation of 162
   collective 163–4, 165–6
   types of 146–7, 171
aggression 114, 116
   see also violence
Ajuriaguerra, J. 47
alienation
   in adaptation 130–1

   of agency 162
   and inequality 165–6
   of labour 153–5
Allport, Gordon 94, 250
alternative life-styles 150
   counter-culture *t. 170–1*
Alternative Ways of Life Project (UNU/GPD) 172
altruism 97, 105–8
analysis, levels of 63–6, 83–6
Anderson, Perry 146
animals 128, 224
anxiety 117–18
Archibald, W.P. 165
Argentina 44
   see also Buenos Aires
Arieti, Silvano 130–1
Aristotle 7, 229
attitudes, social 141–5, 162–3, 177
authoritarianism
   children's experience of 32, 182
   personality mechanism 115
   public, Latin America 193, 198–200, 207–9
   state 12
   work 34

Barker, P. 169, 172
Barreiro, T. 61
Bateson, Gregory 251
behaviour (conduct)
   African 219–21, 223–4, 229
   defending 78–9
   and experience 64–5, 86
   and responsibility 68–9, 79–80
behaviourism 9, 11, 64, 65
   view of the person 66–72, 84
Belgian Congo 22
belief 42, 74–5
Benedict, Ruth 82–3
biological needs 15, 66–7, 72, 76, 78, 87
biology, faith in 84, 85–6

birth 85
body, experience of 30, 41
boredom 49
Bossel, H. 99
brain, development of 47
Braverman, H. 151-2
Britain
  attitudes, social 141-2, 177
  colonialism 22
Browning, H.C., and Singlemann, J. 152
Bruner, Jerome 79
Buddhadāsa Bhikkhu 233, 242
Buddhism 13, 233-47
  the four requisites 234
  the four Sublime Abodes 245-6
  the four wheels 246
Buenos Aires 190, 198, 201, 202, 204, 207-9, 209

Cabral, Amilcar 228
capitalism 64, 65, 67, 229-31, 237-40, 243
  Darwinian 80
  liberal, development model 6-10, 11
  and social character 181
  and work 151-5
Cassirer, Ernst 128
Chang and Eng, Siamese twins 88
change
  readiness for 42
  resistance to 129-30
  social
    strategies for *t. 170-1*
    tendencies *t. 167, 168*
  see also adaptation
character, social 179-88
  and children's needs 182-88
  dynamics 179-2
  growth model 23-4
  transmitted by family 180-3, 184, 186, 188
  transmitted by school 186-7
  see also adaptation, social; personality
children
  Africa 218-19
  early relationships, effects of 31-2, 81, 83, 131-2, 18
  needs 73, 182-6
  primary growth 25-7, 31-2, 47
  psychic growth 27-8, 31
  school, perception of 182, 186-7
  social character transmitted to 180-1, 186-88
    growth model 23-4
  technology, effects of 85
  see also education; family

China, development 234
Christianity
  in Africa 222, 223
  and capitalism 239-40
Clark, Kenneth 86
class, social 165
code, social 128-9
  co-development 98, 107, 108-9, 135
co-evolution 93, 98, 106, 108-9
  field model 102-5
  types of actors in 105-8
  matrix 99-102
  personality type 106
cognitive dissonance 74-5
  uncertainty 135
collective action 193, 197-203, 207, 210
collective needs 203-5, 206, 207, 210
collectivism 5-7, 10-12,16
colonialism, Africa 222, 223, 224, 225-28
  decolonization 226-28
communalism 7, 64, 65
communication 165
  cultural 195-6
  distorting mechanisms of 115-16
    effects on the person 116-18
  in groups of different types 118-19
communism 11, 64, 65, 234, 240
community
  relation 6, 15
  solidarity 202
  see also participation in social decisions; social interaction
competition 8, 187
conditioning 68, 69
conduct see behaviour
consciousness, states of 255-6
  sleep 47, 147-50
consistency, need for 75
  consumerism and consumption 44, 69-70, 230, 258
  advertising 230, 237, 239
  cultural practices 195-7, 202-3
  and needs 192, 204, 207
consumption
control
  agency 146-7
  extrinsic or self- 79-80, 88
co-operation 86
co-operatives 155
  kolkhozes 11
Coopersmith, Stanley 81
counter culture *t. 170-1*
creation
  and cultural practices 195, 196, 197-200, 203

and needs 203-10
creativity 30, 251
crime and violence 131, 141, 142, 159, 183, 239
crises, life 132-3
  nuclear 50-62, *t. 52-3*
  and needs 55-61, *t. 58-9*
criticism 12-13
  social 252-3, 254-5
cultural archaism 159
cultural practices 195-203
cultural sphere, likely tendencies *t. 168*
culture
  African 214-15, 217-32
  counter-cutting *t. 170-1*
  definition 210, 211, 214-15
  maturity growth and 32-8, 38-43
  popular, Latin American 190-211
    academic 209
    consumerist 196-7, 204
    creative-productive 197-200
    intermediate 200-3
    needs associated with 203-5
    social representation and 205-10
  utilitarian 38-43
cultures, different, and normal development 22-4

Darwin, Charles 86
decision-making
  and democracy 162
  in family 155-6
  social *see* participation in social decisions
defending behaviour 78-9
democracy 162, 215-17
  African view of 224-5
  see also participation in social decisions
depression 116, 133
Descartes, René 7
determinism 10, 79-80, 88
  Latin America 193, 207
development
  approaches to 63-5
  continuous 126-8
  idea of 248-1, 256-58
  national 21-3, 43-4, 49, 23
  quantitative and qualitative 233-6, 240-3
  and social adaptation 126-38
  see also human development
deviation and crime 23, 131, 141, 142, 159 183, 239
dignity of man 215-16
dreaming 74, 148-150

Durkheim, Emile 252

economic development 43-4
  quantitative 235-42
economic sphere, likely tendencies *ts. 167, 168*
economics
  classis 7-10
  Gross National Product 236-7, 240
  see also capitalism; industrialism; work
education
  in Africa 225-6, 227-28, 231
  authoritarian 32
  compensatory 47
  for human development 188
  social 23-4
  success in 235
  system 161
  see also schools
ego psychology 77, 96, 250
egoism 105-8
emotions, sensitivity 41
encounter groups 15, 121-2
energy crisis 80
entropy 80-1
environment
  effects of poor 37-8, 49
  natural 17
  and world 127-8
  see also adaptation, social
Enzenberger, M.M. 159
equity 17
  inequality 165
Erikson, E.H. 50-1, 52, 54-5, 56-7, 61
Età Verde (Green Age) 185
European Trade Union Institute 155
evolution theory 66-73, 86
experience
  and behaviour 64-5
  humanist view of 13
  subjective 86-7
externalism (extroversion) 5-7, 9-12, 13, 16

family
  'abnormal' member 132
  African 218-19, 220-1, 225, 230
  child's perception of 182-3, 184
  decision-making in 155-6
  distorting mechanisms in 118
  effects of bad 131-2
  need for 73
  paradigms compared *t. 40*
  parents 30, 31, 32, 198-9

as Primordial primary group 113–15, 123–4
relations 6
as social system 180–1, 186, 188
strategies for change *t. 171*
tendencies, likely *t. 168*
time spent in 164–5
and work 33
see also children; marriage
Fanon, Franz 230–1
Festinger, Leon 75
Fickett, L.P. 64
Fielder, Leslie 88
food, need for 78
see also nutrition
Foucault, Michael 252
France, social character 180
free will 79, 80, 88
Freire, Paulo 227
Freud, Sigmund 14, 77, 78, 123, 129, 148, 150, 251, 252, 253
Fromm, Erich 35, 45, 49, 179, 180, 187, 240, 254, 256
future, the
attitudes to 141–3
children's 182–6
consideration of 17

Gallup, G. 142
Galtung, Johan 65–6, 95, 99, 136
Gandhi, Mahatma 135, 138
Gardiner, W.L. 9
Gerber, Marcelle 85
Gershuny, J.I. 152
and Thomas, G.S. 150, 155, 159
goals
collective, private and public 146
and motives 109
ultimate 234, 243, 245
see also agency
goodness 82–3, 86, 88–9
Gouldner, A.W. 96
Green, H.B. 51, 52, 54
Green Age (Età Verde) 185
green movement 16
groups
community relations 6, 15
distorting mechanisms in 115–19
in different types of group 118–19
effects on the person 116–18
encounter 15, 121–2
healthy 36–7, 48–9, 102, 119–21, 122–3
and individual development 113–15
interaction between 97–8
motivational relations 93–110

primordial primary 113–15, 123–4
psychotherapeutic 121
social, two paradigms *t. 40*
growth *see* development; human development

Harlow, Harry 73
harmony 217–18
health
psychological 4–5, 134–5, 250–3
public 235
Hinde, R.A. 96
Hobbes, Thomas 83
Homans, G.C. 96
Homer, *Iliad* 44, 45
housework 147, 155–6
human development
concept 3–6, 16–17, 61–2
epigenetic ground plans 57–61, *t. 60*
growth 25–30
from inside out 76–7
maturity 26–43
model proposed 23–5
need for theory of 21–5
primary 25–7, 47; and life-stages 61; and maturity 31–2
psychic 27–8
psychological 75
vs. success 39–42
humanist concept of 12–17, 76–7
in industrial societies
life-stages 50–5, *t. 52–3*
and needs theory 57–61, *t. 58–9*
models, dominant 6–12
humanism 24, 45, 64, 65, 66
forms of 14
liberal 7
models of human development 12–17
morality 24
psychology 12–17
view of person 72–83
growth 76–7
intrinsic needs 72–5; hierarchy 75–6
relationships 82–3
responsibility 79–80
worth 80–2
identity
problems 133, 135
security 30
see also adaptation, social; personality; self
ideological systems 3, 5–12, 163
Iliad (Homer) 44, 45
Illich, Ivan 233, 248
India 255, 256, 258

individualism 5, 7-8, 9-11, 13, 16, 215-17
  in utilitarian culture 42-3
individuals *see* adaptation, social
industrial societies, human development in 141-5
  assessing 145-66
    provisional 171
  children in 182-8
  children in 182-8
  taking stock 166-70
  time-budget *t. 149*
  work, conditions and effects 151-7
industrialism 10-11, 12, 145
  and human development 21-2
industrialization, and happiness 142, 235-6
inequality 165
  equity 17
information
  cognitive dissonance 74-5
  communication of 165
  technology 153, 155, 160
  transmitting 159-61
inhumanity 4
institutions
  analysis 63-4
  dominant 166
  participation in 197-200, 203
  social, time use 147, 164
intelligence 26, 28, 47, 48
internalism (introversion) 5-10, 13
internationalism 166
Islam 221-2, 223
Italy
  children 182-6
  industry 174

Jacoby, R. 148
Janov, Arthur 134
Jefferson, Thomas 216
jobs *see* work
Jung, Carl 148

Kalleberg, A.C., and Griffin, C.J. 153
Kamenka, E. 10
Kant, Immanuel 109
Kelley, H.H. 102
Kenya, children in 85
Keynes, John Maynard 239, 240
Kleitman, Nathaniel 74
Knauth, Percy 133
knowledge
  Buddhist concept of 243
  capacity for 29-30
  Latin American concept 209

paradigms compared *t. 41*
  self- 30, 41
Koch, Sigmund 249
Kohn, M.C. 153-4
kolkhozes 11
Kuhn, Thomas 252

labour *see* work
Laing, R.D. 251, 252
Latin America, popular culture study 190-211
leadership 119, 120, 199
  Latin American 208
  psychotherapeutic group 121
  see also authoritarianism
learning 42
leisure activities 157-60, *t. 158*
  Latin American 194-201, 211
Levi, L. 154
Lewin, K. 109
liberal capitalism 6-10, 11
liberal democracy 215-16
liberal humanism 7, 15-16
life-stages 50-5, *t. 52-3*
  and needs theory 57-62, *t. 58-9*
Locke, John 216
longevity 84, 242-49
Luke, Stephen 215

McDougall, W. 94
Maddi, Salvadore 72
madness 250-2
Mallmann, C.A. 94, 95, 145
man, concept of
  African 217-19
  Western 215-17
Mann, M. 163
Marcuse, H. 130, 150
marginated people 13
marriage 128-9, 164
  African 218-19
  children's views of 183, 184
  see also family
Maruyama, M. 94-5. 97
Marxism 10, 14, 217, 229, 253
Maslow, Abraham 13, 60-1, 75, 77, 78, 83, 87, 94, 134, 135, 257, 258
mass media 157-60, 195, 239
material conditions of life 37-8, 49
material needs 43-4
maturity
  concept of 23-4, 45
  growth 26-43
    characteristics 28-30

and culture 32–8
and life-stages 61
primary growth, roots in 31–2
psychic 27–8
and success 38–42
personality 4–5, 134–5
Mazrui, Ali 225
Mbiti, J.S. 218, 219, 222
Mboya, Tom 229
mental development 26, 28, 47, 48
mental illness 250–2
Milgram, Stanley 70
military, the 162
Mill, John Stuart 7, 216
Miller, G. 45
Miller, J.G. 98
modern society 137, 256
children in 187
critics of 13–14
paradigm 12–13
relations in, 6–7, 12–13, 16, 22
see also industrial societies
morality
African 19–1
humanist 24
and motivational field 105–9
public, British 141–2
social code 128–9
motivational fields 93–109
actors in 105–8
balance 98–102
model 102–5
relations
matrix *t. 101*
types 101–2
systems, and distorting mechanisms 116
Mphahlele, Z.K. 224
Munro, Donald 216–17
Murray, H.A. 94
Mussen, Paul 83

Naples 184–5
nations 12
industry and happiness 142
technical development 21–3, 43–4, 49, 233
time-budget *t. 149*
see also industrial societies, state, the
needs
biological 66–7, 72, 76, 78, 87
children's 73, 182–6
collective, associated with cultural practices 203–5, 206, 207, 210
conflict between 77

deficiency and growth 257
for development 126–7, 135
evaluation criteria of 95–8
extrinsic 66–7
hierarchy of 75–6, 77, 87, 88, 94–5, 99
diachronic 57–61
individual 49
intrinsic 72–5
lack of satisfiers 77–9
material 43–4
and motivational relations 93–107
matrix *t. 101*
personal, time spent on 147–51
polar potentialities 55–7, *t. 58–9*
and epigenetic ground plans 57–62, *t. 60*
psychological 73–5, 76
self-actualization 14–15, 82
social character, and satisfaction 180–1
and social interaction 93–107
sociological 72–3, 76, 78, 88
subjective/objective 191–3, 204, 210
neighbourhood associations, Latin American 193, 197–200, 207–8
solidarity 200–2, 203
Nepal, gestures 88
nervous system 66–7, 72, 87
normality
and cultural relativism 22–4
health 250–5
norms, social 128–9, 136–7
social representations, Latin American 205–10
nuclear power 12, 176
Nudler, Thelma 65, 102, 145
nutrition
diet, and women's movement 150–1
early deprivation, effects 31, 47, 49, 131
need for food 78
Nyerere, Julius 223, 224, 227, 228, 229

Oakley, Anne 155

Paci, M. 174
paradox 137
parents *see* family
participation in social decisions 12, 17, 34–5
in Latin America 193–208, 210
and cultural practices 195–203
needs and social representations 203–210
need for 191–3
in work planning 34

Pavlov, Ivan 68
peace movements 166
Pearce, Joseph 85
person, the
  behavioristic concept of 66–72, 84
  conditioned 68
  good 83, 86, 88–9
  humanistic concept of 72–83
  as interchangeable part 70, 81–2
  responsibility 68–9, 79–80
  worth of 69–70, 80–1
personal needs, time spent on 147–51
personality
  African 213–32
  development *see* self
  distorting mechanisms, effects on 116–18
  healthy 4–5, 134–5, 250–3
  types, in motivational field 105–8
  see also adaptation, social; character, social; maturity growth; self
Piaget, Jean 75, 88
pleasure 30
politics 160–4, 166
  tendencies *t. 167*
pollution 17, 242
post-materialists 13–14
potential, actualizing *t. 52*, 76–8, 134
  self-actualization 14–15, 82
potentialities
  creative 30
  polar 55–7, *t. 52, 58–9*
  and epigenetic ground plan 57–61, *t. 60*
  self-actualization 14–15, 82
poverty 190–4, 211
private and public realm 215–17
Protestantism 9
psychic growth 27–8
psychoanalysis 251, 252, 253
  conformist 148, 150
psychological needs 73–5, 76
psychology
  contemporary experience of 249–58
  control 79
  development of 84
  field theory 109
  humanist 13, 14–16
  personality theories 4–5, 134–5
  social 83
psychotherapy 78, 254
  groups 121
public affairs *see* private and public realm; participation in social decisions

quality of life 191–4, 234–5, 241–2
  Buddhist 244–7

Rajneesh, Acharya 254
rationality 12, 29–30, 32, 77, 252, 254
reductionism 65, 87, 128
Reeves, Rosser 44–5
Reimer, Everett 242
relationships 35–6
  community 6, 15
  contractual 71, 82
  distorting mechanisms in 115–16
    effects on the person 116–18
    groups, different types of 118–19
  evaluation criterion of human needs 95–8
  international 12, 17
  intimate/primary 6, 71, 82–3
  in maturity growth 29–30
  in modern society 6–7, 12–13, 16, 22
    motivational 93–109
      matrix *t. 101*
    paradigms compared *t. 40*
    secondary 6
    social 12, *t. 40*
  see also adaptation, social; family; groups; marriage; social interaction
religion 234–5, 238
  African 221–2
repression 129–31, 148, 150
responsibility 68–9, 70, 79
retirement 132
Richards, Audrey 78
rights 215–17
Rome, children of 182–4
Rummel, Rudolf 109
Ryle, Martin 162

São Paulo, Brazil 190, 191, 196–7, 204–6, 208
  leisure 201–2
  schools 198–9
  women 196–7, 200, 201, 205–6
Scheler, Max 244
schools 124, 188
  children's views of 182, 186–7
  São Paulo 198–9
  see also education
Schultz, D. 4–5
Schumacher, E.F. 244–5
science *see* technology and science
security 30, 135
self
  actualization 14–15, 82, 134
  alienating mechanisms 39

development 28, 29, 30
  and groups 113–15
  inner 13
  see also adaptation, social; identity;
    individualism; personality
self-control 79–80, 88
self-esteem 70, 81
self-knowledge 30, 41
sensitivity 30, 41, 135
sensorial-age 51, *t. 53*, 62
sexual impluse 73, 78, 128–9
  see also marriage
Siu, R.G.H. 98
Skinner, B.F. 68, 71
sleep 74, 147–50
social attitudes 141–5, 162–3, 177
social change
  strategies for *t. 170–1*
  tendencies *t. 167, 168*
social character see character, social
social class 165
social code 128–9
social interaction 93–8
  distorting mechanisms of 115–16
    effects on the person 116–18
    in groups of different types 118–19
  and human needs 93, 94–8, 180–1
    see also sociological needs
  motivational balance 98–109
  see also relationships
social norms 128–9, 136–7
  Latin American, representations 205–10
social periphery, urban 190–4, 211
  Latin American 190–211
social psychology 83
social representations, Latin American 207–10
socialism
  African 228–1
  approach to development 64, 65
  Buddhist 246–7
  state-, ideology 6–7, 10–12, 163, 166, 239, 240
  and work 153, 154
socialization see adaptation, social;
  character, social; education
societies, industrial see industrial societies
society
  requirements for human development 16–17
  synergetic or antagonistic 82–3
  see also modern society
sociological needs 72–3, 76, 78, 88, 180–1

evaluation criterion 93, 94–8
solidarity, Latin America 202, 203
Soviet Union 11, 234, 240
Spenner, R.L. 152
Spitz, René 73
sports 158
  Latin America 200–1
Sri Lanka 246
state, the 161–4, 166
  strategies for change *t. 171*
  see also nations
state-socialism see socialism
stimulation, need for 47, 73–4, 75
success and achievement 5, 10, 12, 16, 135
  utilitarian, and maturity 38–40, *t. 40–2*
suffering 254–5, 258
survival, as goal 72–4, 76, 84, 86
suspicion 143–4
symbolism 128, 148
systems
  and entropy 80–1
  hierarchy of 63
Szalai, A. 147, 155, 175

technology and science 12, 21–3, 43–4
  effect on children 187
  effect on work 155
  faith in 85–6
  see also industrialism
television 157–60, 239
Thailand 237, 239–0, 241–2, 246
Thorndike, Edward 68
time
  self in, development 51, *t. 52, 53*
  use 147–65, *t. 149*
    family 164–5
    free 157–60, *t. 158*; Latin America 194–201, 211
    industrial societies *t. 149*
    personal needs 147–51
    state 161–4
    value 242
    work 151–7
Timpanaro, S. 148
Toffler, A. 187
Tolstoy, Leo 133
Touré, Sekou 214
traditionalism 13–14, 254
Truman, Harry S. 248
trust 135, 143–4

Uganda, children 85
uncertainty 135
  cognitive dissonance 75
understanding, humanist view 13, 14

unemployment 156–7, 175
United Nations, Social Development Research Institute 240
United States of America 241, 242
　decision-making, family 155–6
　democracy 225
　social attitudes 49, 142, 163, 177
　work in 153, 154
utilitarian culture 35, 37
　antinomy of 42–3
　company with maturity 38–40, *t. 40–2*

value, of person 69–70, 80–1
value systems 193
　social representations 205–10
violence 159, 183
　aggression 114, 116
　crime 131, 142, 183, 239

war 142, 162
Watson, J.B. 68

Weber, Max 9
women
　children leaving 132
　housework 155–6
　personal care 150
　of São Paulo 196–7, 200, 201, 205–6
　working 151
work
　children's view of 183–4
　degradation of 151–2
　domestic 147, 155–6
　ethics 146
　groups, working 118
　jobs, people and 70, 72, 81–2, 87
　maturity growth and 30, 33–4, *t. 40*
　São Paulo 196–7, 206
　retirement 132
　time spent in 147, *t. 149*, 151–7
world, individual's private 5, 127–8, 133
worth of person 69–70, 80–1
Wright, E.O. 152